Making monoclonals

Front cover Scanning electronmicrographs of NS1 plasmatcytoma cells, spleen lymphocytes, and fusing spleen and NS1 cells

Making monoclonals

A practical beginners' guide to the production and characterization of monoclonal antibodies against bacteria and viruses

Diane G Newell
Brian W McBride
Stuart A Clark

Published by the
Public Health Laboratory Service

Making monoclonals

61 Colindale Avenue, London NW9 5DF

ISBN 0 901144 23 1

Typeset by Quadrant, London NW4
Printed by The Laverham Press, Salisbury

Preface

Many laboratories now find it necessary to produce monoclonal antibodies directed against specified antigens of infectious agents. Such 'designer' antibodies are becoming essential research tools and, more recently, are in demand as immunodiagnostic reagents, the commercial potential of which is enormous.

This book provides the novice with a practical manual containing sufficient information to be able to choose and prepare the antigen, immunize the animals, undertake a fusion and characterize the resulting supernatants.

The content and presentation of this book have been influenced by the recent experiences of the authors during transfers from laboratories where monoclonal antibody production was routine, to environments with less relevant expertise available. Not suprisingly, our initial attempts at making antibodies in these new laboratories were relatively unsuccessful. We have, therefore, incorporated into this book the lessons learnt during that difficult period.

By following the protocols given we believe that the reader will have a good chance of producing monoclonal antibodies at the first attempt. Such successes will, we hope, form the basis for the production of antibodies appropriate for diagnostic purposes by following the relevant literature for improved techniques, by discussion with industrially-based scientists to establish the antibody characteristics most useful in diagnostic kits and by adapting the technology to suit the antibody specificities required.

Contents

Introduction

In August 1975 Georges Kohler and Cesar Milstein published a paper in *Nature* entitled 'Continuous cultures of fused cells secreting antibody of predefined specificity', in which the last sentence stated: 'Such cultures could be valuable for medical and industrial use.' In the intervening years this potential has been fully realised, culminating in Nobel Prizes for the two authors in 1984.

The discovery of antibody-secreting hybrid cell lines originated from experiments designed to investigate the relationship between immunoglobulin (Ig) heavy and light chains during the fusion of myeloma cells of different cell lines but derived from the same inbred mouse strain. This fusion resulted in the production of new Ig molecules consisting of a mixed association between heavy and light chains from the two parental cells. The myeloma cells used in these early experiments secreted Ig molecules of unknown specificity, but subsequently spleen cells from mice immunized with sheep red blood cells (SRBC) were fused with myeloma cells and the resultant cell lines produced anti-SRBC antibodies.

These experiments clearly demonstrated that, despite the inability of terminally differentiated B-lymphocytes to grow in culture, the fusion of these cells with myeloma cells can produce hybrids which are capable of long term culture. Moreover, these hybrids produced Ig molecules bearing the variable domains determined by the immune B-lymphocyte.

This artificial generation of immortalized antibody-producing cells is the basis of monoclonal antibody technology.

In the simplest of terms the actual procedure of monoclonal antibody production involves the fusion of an immune spleen cell with a myeloma cell in the presence of polyethylene glycol (PEG). The resulting hybrid is grown in a selective medium (HAT medium; see page 16) which precludes the growth of non-hybrid cells. In order to aid selection the hybrids are cultured in multi-well plates so that only few hybrids will be seeded into each well. Within 7-10 days each hybrid will have grown into a small, but visible, clone actively secreting immunoglobulin into the culture supernatant. These supernatants are then tested for antibody activity by use of a suitable screening assay. Cells from those wells with positive supernatants are then removed and cloned in order to ensure that only strongly growing and antibody-producing cells are present. As early as possible some of the cells are stored in liquid nitrogen for security. After the cloning procedure the cells are bulk cultured to provide antibody-containing supernatant. Alternatively, if large quantities of antibody are required, the cells are injected into the peritoneal cavity of syngeneic mice where they grow as a tumour and induce the production of highly concentrated antibody in an ascitic fluid. Finally the properties of the antibody, especially its antigenic specificity, need to be identified; this

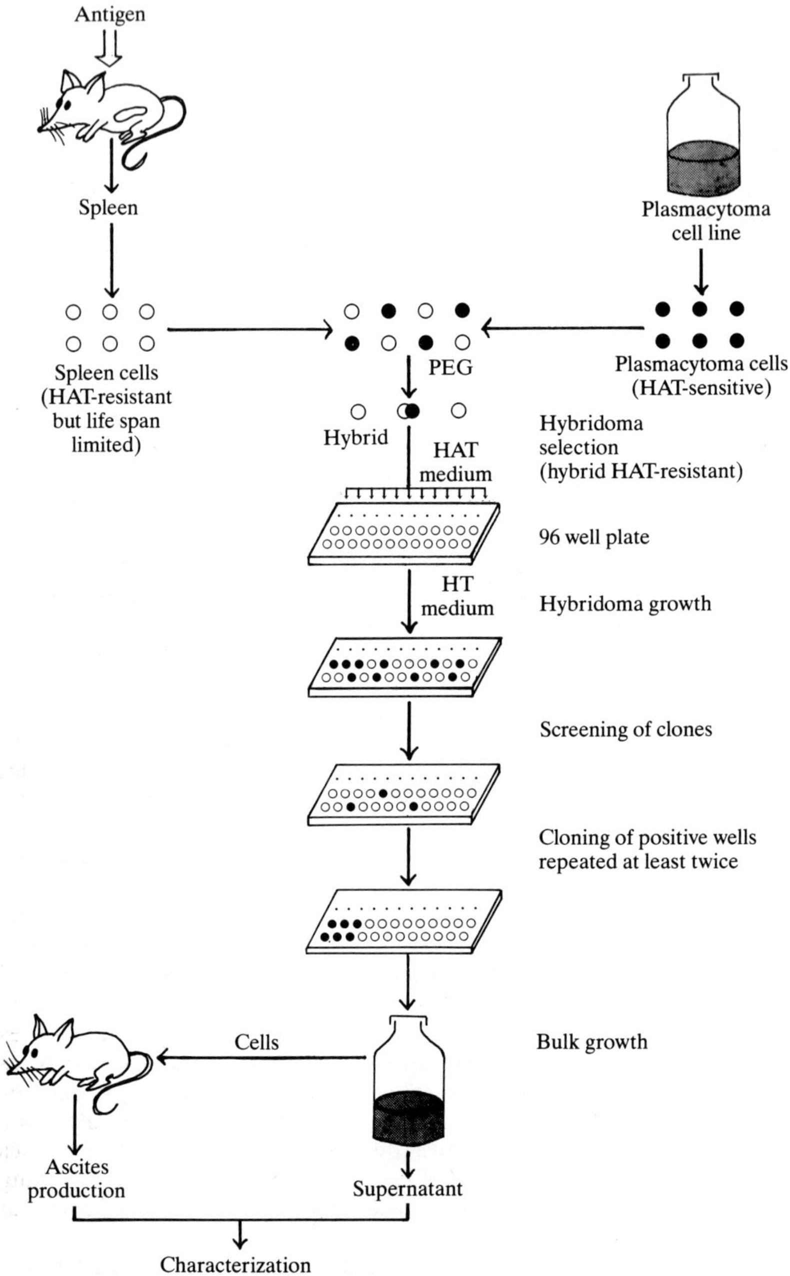

Figure 1 Scheme for monoclonal antibody production

process of characterization will determine its potential usefulness as a diagnostic tool.

Since the original experiments, the techniques for hybridoma production have been so refined that the technology is well within the capacity of even laboratories possessing only modest tissue-culture facilities. But before embarking on the production of monoclonal antibodies for the diagnosis of infectious disease, three key questions should be asked.

- Is it feasible and advantagous to diagnose the disease by use of immunodetection techniques?
- What advantages will a monoclonal antibody have over already available polyclonal antisera?
- What will be the cost in terms of finance and resources?

The first question will, of course, need to be considered for each infectious disease. Diseases in which the infective agent is not culturable, such as Norwalk virus, or slow growing, such as *Campylobacter jejuni*, are prime candidates for immunodetection techniques. In other cases it can be difficult to see the advantage in developing an immunodiagnostic kit.

The second question would also appear to be easily answered. Monoclonal antibodies are directed against single antigenic determinants and are, therefore, uniquely specific for those determinants, unlike the multiple specificity of polyclonal antisera. And a suitable hybridoma can provide an infinite source of standardized monoclonal antibody, while hyperimmunized animals require frequent replacement. The antigenic specificity is not necessarily absolute, however; common antigenic determinants, such as the carbohydrate groups of glycoproteins and glycolipids, may be detected. Moreover, this specificity can be a distinct disadvantage when genus or species-specific, rather than strain-specific determinants are to be detected.

Third, the time and costs involved in the production of monoclonal antibodies are undoubtedly far higher that those required to produce conventional antisera. It is assumed, however, that laboratories already engaged in tissue culture techniques will have most of the facilities necessary. Of course, the production of monoclonal antibodies directed against minor or poorly antigenic determinants may require some innovation and extra time, but the commercial rewards for such antibodies can be significant. Nevertheless, should suitable monoclonal antibodies be available from other laboratories, the effort involved in making new antibodies is unlikely to be worthwhile. It must be emphasized that once the fusion has been successfully achieved, the work load increases alarmingly. The rapid screening of supernatants and subsequent cloning of positive cultures will consume considerable and continuous effort for at least 10 weeks.

The final, and possibly most important consideration is the detection system for any antibodies produced. A suitable screening assay is, therefore, an essential prerequisite to any attempt at fusions.

Chapter 1

Immunization

Thousands of antigens may be expressed on the surface of a virally infected cell or a bacterium. The ultimate advantage of monoclonal antibody technology is that specific antibodies may be obtained despite immunization with such a multiplicity of antigenic determinants. In practice, the use of such complex antigens necessitates considerable effort in the selection and characterization of the monoclonal antibodies produced. Moreover, unwanted antigens may dominate the immune response. Therefore, enormous advantage can be gained from purification, or even partial purification, of the desired antigen. Nevertheless, it must be remembered that purification may expose epitopes not normally seen in the antigen's native state, while other epitopes may be destroyed. It is, therefore, essential to check the immunological integrity of the purified antigenic material, prior to immunization, using the desired screening system.

The considerable variation in the properties of bacterial and viral antigens precludes any attempt to describe specific preparative procedures in detail. Some methods for antigen preparation are given, but suitable techniques will need to be developed for each organism. Although no attempt has been made to describe techniques for parasitic or fungal organisms, many of the comments and procedures described below may be extrapolated to other infectious agents.

Immunization procedures

In the derivation of monoclonal antibodies the primary aim of immunization is to stimulate the production of clones of antigen-committed B-lymphocytes, to a point of differentiation which is suitable for fusion. A significantly higher number of specifically stimulated B-cells will be obtained if the antigen is presented optimally. Antigens are, in general, much more immunogenic when presented in an insoluble or particulate form or when suspended in an adjuvant.

It must be noted that the immunization of animals is now subject to

Home Office Regulations, and Personal and Project Vivisection Licences are required for this type of work.

Mouse strains for immunization

The plasmacytoma cell lines most commonly used for hybridoma production are derived from the inbred mouse strain BALB/c (see page 18). Although mixed strain, and even mixed species hybridomas can be produced, the fusion products tend to be less stable and to expel chromosomes at random. Moreover, the bulk production of monoclonal antibodies by means of ascitic fluid with mixed hybridomas would be very difficult (see page 45). In practice, hyperimmunized BALB/c mice are the usual source of the parental spleen cells. BALB/c mice are readily available and easy to handle, but other inbred mouse strains can mediate greater immune responses, especially with poor immunogens. When an enhanced antibody response is required, inbred mice which are the first generation of a BALB/c x CBA mating (F1(BALB/c x CBA)) may give better results.

Adjuvants

The most common adjuvant in use is Freund's, which is available in complete (CFA) and incomplete (IFA) forms. Freund's complete adjuvant is a mixture of oil (Bayol F) and detergent (mannide monooleate) containing *Mycobacterium tuberculosis*. The incomplete form comprises oil and detergent alone. It should be emphasized that these adjuvants must be prepared with the greatest care, since accidental injection into the hand may result in permanently stiff fingers or very severe hypersensitivity reactions (Chaple & August 1976). Care should also be taken to avoid eye contamination as this may also result in permanent damage.

Other adjuvants include the addition of *Bordetella pertussis* organisms to alum-precipitated proteins, aluminium phosphate and aluminium hydroxide. These may also be used in conjunction with Freund's adjuvants.

Preparation of Freund's adjuvants

1 Shake the vial of CFA thoroughly to re-suspend the mycobacteria.

2 Take up the aqueous antigen solution into a 5 ml glass syringe fitted with a Luer lock. (The plungers of plastics syringes swell in Freund's adjuvants.) A second syringe is filled with the correct volume of adjuvant. A ratio of 1:1 adjuvant : aqueous antigen is generally recommended, but ratios of 2-4:1 may be used (Hurn & Chatter 1980). Because of the significant fluid loss in the syringes and needle during this process it is recommended that twice the required volume be emulisified.

3 Expel any air from the syringes and fit a double metal hubbed needle of 1mm diameter or a three-way plastics stopcock (Viggo AB, Helsingborg, Sweden).

4 The aqueous phase is injected into the oil and the mixture is passed rapidly back and forth until the emulsion is thick and creamy. The presence of detergents in the antigen mixture will prevent full emulsification and the emulsion will be less stable. The emulsion is tested by placing one drop onto the surface of some water, where it should remain as a discrete globule.
5 Before injection ensure that the emulsion will pass through the needle. The emulsion should be stable at 4°C for several weeks, but must not be frozen.

Alum precipitation of antigens
1 To 10ml of protein solution add 4.5ml of 1M sodium bicarbonate.
2 Slowly add 10ml of 0.2M aluminium potassium sulphate while stirring. Leave for 15min.
3 Spin at 300g for 15min and wash the precipitate three times with phosphate buffered saline (PBS).
4 Resuspend the insoluble material to the required concentration (approximately 4mg/ml).
5 If required mix 0.1ml of the antigen with 4×10^9 *B pertussis* (0.1ml of whooping cough vaccine).

Routes and timing of injections

Rodents may be injected intradermally (id), intramuscularly (im), subcutaneously (sc) or intraperitoneally (ip) with CFA, but it must never be given intravenously (iv). The first injection of antigen should be given in CFA, while subsequent injections are given in IFA in order to avoid ulcer formation.

Occasionally soluble aqueous antigens are given ip or iv throughout the immunization schedule, but these routes tend to stimulate poor immune responses and to produce a low frequency of specific antibody-secreting hybrids.

The timing and number of the immunizing boosts may make a significant difference to the immune response. A reasonable circulating antibody response may be detectable with just a single priming injection of antigen. However, this is likely to comprise a primary response only and therefore to mediate mainly IgM-producing hybrids, while longer schedules, with multiple boosts, tend to give, at least with T-dependent antigens, antibodies of the IgG and possibly IgA classes. If time and antigen are precious, then intrasplenic immunization may be considered (see opposite).

Recommended schedule of immunization

Inject a total of 5-50μg of protein emulsified in CFA (1:1 ratio) into four sites sc at the scruff of the neck (0.1ml/site) using a 23g needle and a glass syringe. (Higher antigen doses may lead to the production of anaphylactic shock on boosting.) Within 2 weeks, the mouse should have developed a number of small granulomas at the sites of the injections. Two to three subsequent injections of the same amount of antigen emulsified in IFA are given at 10-14 day intervals by the same route. The mouse is bled either by cardiac puncture (see page 26) or from the tail, by excising the tip with sterile scissors and collecting the blood into a capillary tube. The circulating antibody level is determined by the appropriate screening technique. Once the mouse has generated a sufficient immune response, a final boost of antigen in isotonic solution is given iv, 3 days prior to harvesting the spleen for fusion. Intravenous injection is most efficiently achieved by gently warming the tail of the mouse with a hair dryer or desk lamp to produce dilation of the tail veins. The mouse should be immobilized by anaesthesia or in the barrel of a 50ml plastic syringe (a slot cut in the syringe barrel so that the tail can be exposed). The antigen solution (0.1ml) is then injected with a 25g needle about 1cm from the base of the tail. If the procedure is successful, the tail vein will blanch briefly and the fluid should be seen entering the vein with the minimum of pressure being required on the plunger. Blanching of the tissue around the vein indicates subcutaneous injection, and another vein should be tried.

Whole bacteria and infected cells may be injected without adjuvants. Cells should be washed in PBS to remove media components and 10^7 cells in 0.5ml PBS should be injected ip. Recipients are boosted up to three times at intervals of 4 weeks and fusion is performed 3 days after the last boost, which should also be given ip.

Intrasplenic immunization

Intrasplenic immunization requires only small amounts of antigen and can be performed as a single step (Spitz 1986).

1 Anaesthetize the mouse by ip injection of 10μl/g body weight of the mixture:
Nembutal 40mg
saline 8ml
absolute ethanol 2ml.

2 Place the mouse on its right side, clip the fur and wash the skin with 70% ethanol.

3 Make an oblique skin incision 1-1.5cm long just below the left set of ribs, and separate the skin from the muscle layer by blunt dissection.

4 Pick up the muscle layer and peritoneum with forceps and make an incision carefully, avoiding the intestines.

5 Visually locate the spleen, adjacent to the stomach, and exteriorize by gently lifting the lower pole with sterile forceps.

6 Insert a 27g × ½in needle fitted to a 1ml syringe and inject 50-100μl of antigen solution (200μg/ml) while withdrawing the needle. For immunization with whole cells, 2.5×10^5 cells, in the same volume, are injected.

7 Gently push the spleen back into the peritoneal cavity and suture the peritoneum and muscle layers. The skin edges are closed with three stitches.
8 Remove the spleen 3 days later for fusion.

Viral antigens

The preparation of viral antigens presents special problems, particularly when the virus remains cell-associated in culture—resulting in considerable contamination of the virus preparation with host proteins. These proteins will stimulate the production of antibodies which, if the same antigen is being used for both immunization and screening, will generate irrelevant positive results. There are several ways of overcoming this problem. The antigen used for the screening assay can be from infected cells of a different species. However, this will not eliminate all false positive results, as many of the antibodies will be cross-reactive. Moreover, viral proteins, especially glycoproteins, may not be of the same composition when expressed on the screening cells as those expressed on the cells used for the preparation of the immunization antigen (Hirsch *et al* 1981). Alternatively, the screening assay may include a non-infected cell antigen in addition to the infected cell antigen, but this will obviously double the work load. It is advisable, therefore, to immunize with the purest viral antigen available.

Viral antigens can be divided into two general categories—viral particle antigens and soluble antigens (some of which will be present in the viral particle antigens).

Soluble antigens are usually purified from infected cell-culture supernatants by standard protein fractionation methods (Lennette & Schmidt 1979). Certain viruses, such as some strains of herpes simplex, will release fully mature forms of the virus into the tissue culture supernatant, which can then be purified by concentration and ultracentrifugation on density gradients (Pertoft 1980).

In a few cases mice may be infected directly with the virus. Infection with some togaviruses, for example, will lead to death in newborn but not adult mice. Antigenic material can be obtained, therefore, by use of tissue extracts prepared from infected syngeneic infant mice.

The methods available for production and purification of viruses and their subunits are too numerous to be listed in this book. The reader is advised to consult references relevant to the required virus. When virus purification is undertaken, however, it should remembered that extraction of viral components from cells by detergent treatment will undoubtedly affect the antigenicity of the native protein, and screening should be carried out on similarly prepared material.

Adjuvants for immunization with viral antigens

Freund's adjuvant is useful for increasing the antigenicity of some viruses—for example myxoviruses (Himmelweil 1960)—but is reported to be variable for others, such as picornaviruses and adenoviruses (Tent & Stone 1961). Viral particles may be absorbed onto aluminium phosphate, which will also tend to concentrate and purify the preparation. This preparation can then be used with CFA.

Inactivation of virus preparations

In many cases, preparations of live viruses can be injected ip, in an adjuvant emulsion, into adult mice, without adverse risk to the animal. Under some circumstances, however, (eg a hazardous virus or a susceptible mouse strain), inactivation of the virus may be necessary. Although viral preparations are usually inactivated with chemical agents, some viruses can be readily attenuated by passage in cell culture, but care must be taken to ensure that this procedure is not accompanied by alterations in the antigenicity of the required proteins.

Ideally, the agents used to inactivate viruses should affect the viral nucleic acid without changing the immunogenicity of the capsid or envelope proteins.

Formalin, although frequently used, cross-links proteins, which results in denaturation, as well as reacting with the amino groups of nucleotides. For vaccine production, formalin is usually used at a dilution of 1:5,000 but the inactivation time may be prolonged—for example, polioviruses should be treated for 7-15 days at 37°C. While the reaction will follow first-order kinetics for purified viral preparations, its effects in crude preparations will be more complex.

Glutaraldehyde, 0.1% (v/v) may also be used for inactivation, but, as with formalin, the time of exposure and temperature at which inactivation is optimal while leaving the sample immunogenic, must be determined empirically.

ß-propiolactone is a very effective agent for the inactivation of some viruses, but is known to alter the antigenicity of certain viral proteins—for example, the haemagglutinin of influenza viruses (Goldstein 1970). ß-propiolactone may be used at a final concentration of 0.025% (v/v) and the sample treated at 4°C for 20h followed by 2h at 37°C.

Other chemical agents at present being used for the production of viral vaccines include ethyleneimine, propyleneimine and N-acetylethyleneimine (Morgan *et al* 1980)

Psoralen derivatives, such as 4′-hydroxymethyltrioxsalen and 4′-aminomethyltrioxsalen, bind covalently to the bases of nucleic acids in a photochemical reaction requiring irradiation with ultraviolet light. These chemicals have been successfully used for the inactivation of influenza and herpes simplex viruses (Redfield *et al* 1981).

Ultraviolet light at 260nm inactivates viruses, but can lead to multiplicity reactivation in which the genomes from two or more inactivated viruses co-operate to produce infectious particles. It also has little penetrating power and produces areas of shadow in particulate matter. The aqueous sample, with a liquid depth of no greater than 5mm, should be exposed in an uncovered plastic petri dish. The conditions for inactivation will depend on the ultraviolet source, but 5min at 30cm from the source is usually sufficient.

Ultraviolet light at 235nm is proteotropic and may alter the antigenicity of the proteins; it is not recommended.

Bacterial antigens

Because of the complex antigenic nature of the bacterium it is recommended that the mouse is immunized with as pure an antigen as possible. However, in certain circumstances whole bacterial cells need to be used.

Whole bacterial cells

Lipopolysaccharide (LPS) is such an immunodominant surface antigen that immunization with whole bacteria will generate primarily anti-LPS antibodies.

Whole bacteria are usually washed, to remove the culture medium, and killed before injection. There are four common methods of killing:

- Heat treatment at 60°C for 10min significantly reduces the viability of most bacteria while destroying none but the most heat-labile surface antigens.
- Bacteria may be chemically fixed with 0.1% (v/v) buffered formalin, for at least 1h, or 0.1% (v/v) glutaraldehyde in 0.1M phosphate buffer, pH 7.2, for 10min at room temperature. Aldehyde fixatives cross-link proteins, significantly altering the three-dimensional configurations, and monoclonal antibodies against such structures may not recognize the native antigens. The chemical agents 0.1% (w/v) sodium azide and 0.01% (w/v) thimerosal may also be used. Chemical agents must be removed from the killed bacteria before injection, by either washing or dialysis.
- ^{60}Co irradiation of pathogenic bacteria is an excellent means of retaining fragile antigenic structures while eliminating the viability of the bacteria.
- Mechanical disruption of bacteria will significantly reduce viability and is a prerequisite step to the preparation of outer membranous material. Sonication is the most frequently used method. The bacteria are sonicated at 10mA for 30s bursts interspersed with 30s rest periods. It is usually sufficient to repeat this cycle three times. In order to reduce heating effects the material for sonication should be held in an ice bath. The French Press is usually more efficient when the bacteria are resistant to sonication. Intact bacteria can be removed by centrifugation. Of course cellular disruption releases intracellular material, which may be highly antigenic and contain intracellular proteases.

Cell walls

Bacteria are mechanically disrupted by sonication. The insoluble cell envelope is separated from the intercellular contents by centrifugation and, if required, the membranes are solubilized with detergents.

Outer membranes

The bacterial outer membrane, by virtue of its early presentation to the immunocompetent cells of the host, will be antigenically very important.

Two major methods are available for the preparation of outer membranes from Gram-negative bacteria. These procedures may result in the loss of peripheral membrane proteins or expose antigens at the inner surface of the outer membrane.

Spheroplast formation Osmotically sensitive spheroplasts are produced in the presence of lysozyme (10mg/ml) after the slow addition of ethylenediamine tetra acetic acid (EDTA, 4.5mM, pH 7.5), to give a final dilution of 3mM EDTA. The cells are then gently stirred at room temperature and spheroplast formation is monitored by phase-constrast microscopy. These spheroplasts, when placed in three volumes of ice-cold distilled water containing 0.2% (w/v) Brij 58, undergo osmotic shock and lyse. Separation of the membranes is enhanced by incubation with 14 units/ml DNAase 1, 15 units/ml RNAase and 3mM magnesium chloride for 20 min at 37°C. The inner and outer membranes are then separated on a sucrose gradient (Miura & Mizushima 1968, Osborn & Munson 1974).

Sodium N-lauryl sarcosine extraction of membranes Crude membranes are prepared from sonicated bacteria by high-speed centrifugation (100,000g for 2h). The inner membrane is selectively solubilized from crude membrane by the ionic detergent 1% (w/v) sodium N-lauryl sarcosine (sarkosyl) in 7mM EDTA, pH 7.6, for 20min at 37°C, leaving the outer membrane which is recovered by centrifugation (Filip *et al* 1973).

In Gram-positive bacteria the digestion of the peptidoglycan layer with lysozyme produces an osmotically fragile protoplast which is bounded by the cytoplasmic membrane. Osmotic lysis of these protoplasts will give membrane vesicles which can be recovered by centrifugation.

Flagella

Although flagella are a major contaminant of outer membrane preparations, they may be purified independently by mechanical shearing from whole bacteria in a laboratory blender. Cellular debris is removed by centrifugation and the flagella are isolated from the supernatant by centrifugation at 100,000g for 2h. Flagella dissociate into subunits at pH 2.0-3.0. Reassociation may occur at neutral pH and may be encouraged by seeding with small amounts of native flagella or precipitated with ammonium sulphate (Silverman & Simon 1977). Antigenic preparations containing high proportions of flagellin may also be obtained by extraction of whole flagellated bacteria in 0.2M glycine buffer, pH 2.2.

Polymeric bacterial flagellin is a T-independent antigen and, therefore, will frequently stimulate the production of IgM antibodies despite long-term immunization regimens.

Fimbriae

Fimbriae can be mechanically removed from cell surfaces or dissociated from some intact bacteria by extraction in alkaline conditions (Virgi & Heckels 1985). In the latter case, the pili are disaggregated into subunits and the intact bacteria are removed by centrifugation. Pili are then precipitated by ammonium sulphate.

Lipopolysaccharide

LPS is easy to purify but difficult to exclude from other antigen preparations. The common method of preparation is by extraction of intact bacteria (or, preferably, cell wall preparations) with 45% (w/v) phenol at 65°C. The solubilized LPS is then recovered in the upper aqueous phase on centrifugation at 0°C (Jann & Westphal 1975).

Purified LPS induces antibody reponses mainly against core determinants. Because these determinants are highly conserved these antibodies tend to be very cross-reactive, whereas monoclonal antibodies against LPS, with whole bacteria used as the immunogen, are usually serotype specific.

The repetitive identical antigenic determinants on LPS stimulate a strong but predominantly IgM response.

Extracellular polysaccharides

Bacterial capsules and slimes comprise negatively charged acidic polysaccharides secreted outside the cell wall, and are sometimes found in the culture supernatant (Jann & Westphal 1975). These polysaccharides are poorly antigenic in a purified state, and anti-capsular antibodies are usually obtained by immunization with viable whole organisms. Formalin fixation will result in loss of these antigens.

Purification of the extracellular polysaccharides is possible but requires very harsh chemical procedures. Moreover, adequate antigenicity can be obtained only by coupling to a protein carrier such as bovine serum albumin (BSA).

Extracellular protein antigens

Many proteins, including enzymes and toxins, will be secreted into the culture supernatant. Like many of the previously described antigens, the culture conditions may substantially influence the production and secretion of these proteins. The culture supernatant should be concentrated and then the extracellular proteins should be purified from the media components. Chemically defined media allow easier purification but may not provide the optimal conditions for production. Many of these extracellular antigens will be susceptible to proteolysis and need to be protected by protease inhibitors, which, of course, should be removed before immunization.

Both toxins and enzymes can cause problems when used to immunize animals.

Toxins

Many toxins, such as tetanus and botulinum toxins, are lethal to mice. Detoxification can be achieved by dialysis against 0.4% formalin in PBS for 24h at room temperature (Sato 1987, Oguma *et al* 1986), or toxin fragments may be used. Obviously, such procedures can radically affect the immunogenicity of the toxin.

Enterotoxins do not usually require detoxification before use as immunogens; for example, up to 20μg of purified *Clostridium perfringens* enterotoxin may be injected sc in adjuvant (Tranter 1987). The enterotoxins of *Vibrio cholerae* and *Escherichia coli* may also be injected in their native form (Svennerhol et al 1987).

Despite the lethality of neurotoxins such as botulinum toxin A and B, it may be necessary to immunize the mouse with purified, native toxin. It is essential in this case to start at a very low concentration of antigen. This concentration is then gradually increased during the immunization schedule (Shone *et al* 1985). If necessary the mouse is protected at the primary immunization by concomitant passive immunization with anti-toxin antiserum (Haberman & Gortzki 1985; Thompson *et al* 1986). This schedule permits the mouse to generate a neutralizing response against the toxin prior to encountering it in an active form.

Most recently *in vitro* immunization procedures have been developed which (see page 14) may avoid the lethality of neurotoxins in the whole animal.

Infections

Many pathogenic bacteria and parasites will infect mice, inducing an immune response mimicking the naturally acquired disease. This is particularly important if mucosal immune responses are required. Boosting with pure antigen or *in vitro* stimulation will maximize the chance of getting immune cells at the correct stage of differentiation. Recovery of lymphocytes from sites other than the spleen—for example, Peyer's patches or lymph nodes—may be necessary. It must be noted that infection can induce an immune response of a different isotype to that obtained by experimental immunization (Handman & Remington 1980, Johnson *et al* 1981).

In vitro immunization

Considerable advantages can be gained from the immunization of non-immune spleen B-cells *in vitro*. The amount of antigen required is significantly smaller and the immunization time shorter than when the whole animal is immunized. Moreover, certain biological properties of the antigen disadvantageous to the whole animal, such as neurotoxic activity, may be overcome.

In principle, immunocompetent B-lymphocytes isolated from a non-immunized mouse are incubated in the presence of antigen. The antigen is pre-incubated with macrophages which process the antigen and present it to helper T-cells. In the presence of these antigen-specific T-cells and various growth factors, there is a clonal proliferation of B-cells expressing antigen-specific surface immunoglobulin.

Preparation of antigens

The antigens used in this procedure may be soluble, particulate or whole cells. All antigens to which the isolated immune cells will be exposed should be sterile and free of materials which may be toxic to the cells. Irradiation (4500 rad ^{137}Cs), or treatment overnight in 95% ethanol at 4°C followed by washing three times, is sufficient to inhibit proliferation of bacteria or cells, without severely damaging the antigens expressed on their surface (Reading 1976).

Conditioned medium

In order to initiate an immune response *in vitro*, it is necessary to provide the culture with a rich mixture of soluble factors, such as lymphokines. These factors are secreted by isolated thymocytes into the medium during culture. This thymocyte-conditioned medium (TCM) can then be added to the medium in which the spleen cells are cultured for *in vitro* stimulation. Although rat TCM may be used in murine *in vitro* stimulation experiments, murine TCM is more commonly prepared from mixed allogeneic murine thymocyte cultures.

Materials

RPMI 1640 medium plus 20% heat inactivated fetal calf serum (FCS), HT supplement (see page 23) and 5×10^{-5}M 2-mercaptoethanol
6 BALB/c and 6 CBA mice
Sterile tea strainer or stainless steel mesh cut into 2.5cm squares (Expanded Metal Co, Hartlepool; Type 990, 200mm wide)

Method

1 The 12 mice are sacrificed by cervical dislocation, and the thymuses are asceptically removed (see page 24) and immediately placed in medium.

2 The tea strainer or grids are placed in a sterile 60mm petri dish. The thymuses and medium are poured into the strainer and then forced through the metal mesh with the barrel of a sterile 10ml plastics syringe.
3 The resultant cell suspension is left to stand for 5min in a sterile plastics tube to allow the larger tissue fragments to settle out. The upper single cell suspension is retained.
4 A cell suspension of 4×10^6 cells/ml is then cultured as 50ml aliquots in $75cm^2$ flasks for 48h.
5 After culture the medium is centrifuged at 200g to remove the cells and filtered through a $0.22\mu m$ filter. The filtrate is stored at $-50°C$ or $-70°C$, in 20ml aliquots.

Procedure for *in vitro* stimulation

Materials

RPMI 1640 medium containing 20% heat-inactivated FCS, 10% TCM, $5 \times 10^{-5}M$ 2-mercaptoethanol and HT supplement
1-$10\mu g$/ml soluble antigen in medium
Peritoneal macrophage feeder layers in 24-well plates (see page 24)

Method

1 Macrophage feeder layers are incubated with 1ml of medium containing 1-$10\mu g$/ml soluble antigen for 4h. The medium is replaced with 1ml fresh medium.
2 A spleen cell suspension from an unimmunized BALB/c mouse is prepared (see page 26).
3 1ml of spleen cells (1×10^6) is added to each of the wells containing antigen-primed macrophages.
4 The cells are cultured for 5-7 days and then used instead of the immune spleen cells in the fusion (see page 26).

In vitro immune stimulation is still under development, and several procedures have been suggested, which differ particularly in the composition of the medium (Reading 1976, Borrebaeck & Moller 1986, Boss 1986).

Chapter 2

Fusion techniques

The choice of plasmacytoma cell lines is critical for a successful fusion. In order to ensure that the hybrid cells are not outgrown by the myeloma cells, it is necessary in practice to include a selection procedure. For this purpose, cell lines have been derived which are defective in a salvage pathway enzyme involved in DNA synthesis. The most commonly used myelomas lack the enzyme hypoxanthine-guanine phosphoribosyl transferase (HGPRT). These enzyme defective variants were selected by the inclusion of 8-azaguanine (8-Ag) in the medium.

Aminopterin is included in the medium for hybridoma growth, so that the main DNA synthesis pathway is blocked, by inhibition of the enzyme dihydrofolate reductase (Littlefield 1964). Only those hybrid cells which have a functional HGRPT enzyme, inherited from the spleen parent cell, will survive in culture. Hypoxanthine and thymidine are also included in the medium (HAT medium) as sources of exogenous nucleotides enabling the rescue pathway to function efficiently. Thus only a successful fusion product will survive in HAT medium: the spleen parent cell provides the functional enzyme required to circumvent the aminopterin block, while the plasmacytoma parent provides the ability to grow *in vitro* (Figure 2).

Plasmacytoma fusion partners

The plasmacytoma cell line originally used, P3-X63-Ag8, secretes an IgG_1 kappa myeloma protein called MOPC 21. Many of the hybridomas generated from this plasmacytoma also secrete immunoglobulin containing the heavy or light chains of the MOPC 21 protein. To increase the probability of producing antibody-secreting hybridomas, sublines defective in Ig secretion were selected. Currently, four murine plasmacytoma cell lines are commonly used in fusion experiments (see Table 1, page 18). NS1 produces but does not secrete *k* light chains. So some of the hybridomas produced with this cell line will synthesize MOPC 21 *k* light chains. NS0 and 653 are non-Ig-producing sublines derived from P3. SP2/0 is derived from a murine hybridoma which secreted monoclonal

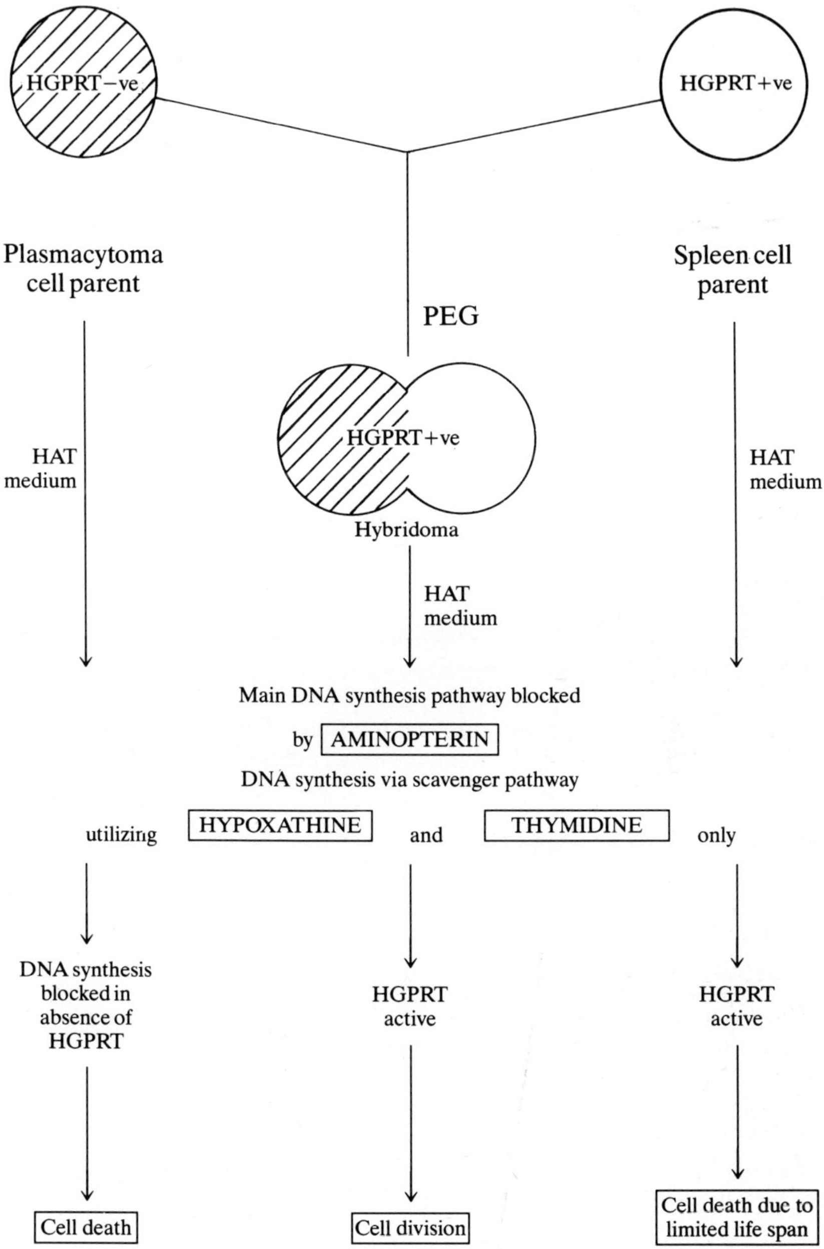

Figure 2 Selection of hybridoma cells

Table 1 Plasmacytoma fusion partners

Complete name	Short name	Mouse/rat strain origin	Ig produced	Reference
P3-X63-Ag8	X63	BALB/c mouse	IgG1,κ	Kohler & Milstein (1975)
P3-NS1-Ag4-1	NS1	BALB/c mouse	κ*	Kohler & Milstein (1976)
NS0	NS0	BALB/c mouse	none	Galfre & Milstein (1981)
P3-X63-Ag8-653	653	BALB/c mouse	none	Galfre & Milstein (1981)
SP2/0-Ag14	SP2/0	BALB/c mouse	none	Shulman *et al* (1978)
210.RCY3.Ag123	Y3	Lou rat	κ	Galfre *et al* (1979)
YB2/0	Y0	Lou/AD rat	none	Kilmartin *et al* (1982)

*κ light chain not secreted

anti-sheep red blood cell antibody. Although SP2/0 tends to have a very high fusion frequency rate, the high chromosome number generated tends to make hybridomas less stable than those derived from other parental plasmacytomas.

In some instances—for example, when large quantities of ascitic fluid or anti-murine antibodies are required—rats are used as the immunized animal. The rat plasmacytoma parental cell lines most frequently used in these fusions have been derived from the Lou rat myeloma 210 (Table 1). It should be noted that Lou rats may carry Hantaan virus and present an infection risk. So prescreened rats are recommended. Alternatively, animals may be tested for Hantaan virus by arrangement with the Special Pathogens Group, Pathology Division, CAMR, Porton Down, Salisbury, UK. There is, apparently, no evidence of Hantaan virus in Y3 or Y0 myeloma cell lines.

Whichever plasmacytoma cell line is chosen for fusion experiments, the source is extremely important. It is essential that the cells be from a reputable laboratory (see page 59) and have a proven record of successful fusions. And it is wise, on receipt of a new cell line, to check for both HAT sensitivity and mycoplasma contamination.

Checking for HAT sensitivity

The HAT-sensitive plasmacytoma cells should die in HAT medium but survive in 8-Ag or 6-thioguanine (6-Tg). To check that the cells have not reverted, one dilution cloning plate of cells is set up (see page 31) in HAT medium and a second plate in medium containing 20μg/ml of 8-Ag or 6-Tg. If cells survive in the HAT medium then a clone, which retains the defective HGPRT enzyme, should be reselected from the 8-Ag or 6-Tg-containing plate.

Checking for mycoplasma contamination

Before any newly acquired cell line is introduced into a laboratory producing hybridomas, it is essential that it be quarantined from all other cells until proven to be free of mycoplasma contamination. As mycoplasma use the supplements hypoxanthine (H) and thymidine (T) for DNA synthesis, and HT supplements are necessary for the growth of hybrid cells, mycoplasma contamination results in poor fusion rates and causes early hybridoma mortality. The European Collection of Animal Cell Cultures (see page 59) will undertake tests for mycoplasma contamination of plasmacytomas or hybridoma cells.

The recommended procedure for the identification of mycoplasma contamination uses the fluorochrome Hoechst 33258, which specifically binds to DNA. The growth of contaminating mycoplasma is enhanced in the presence of emetine, which inhibits eukaryotic protein synthesis.

Materials

10μg/ml Hoechst 33258 in PBS (see page 50)
1μg/ml emetine hydrochloride in antibiotic-free tissue culture medium

Method

1 Culture the cells in antibiotic-free medium containing emetine for 48h.
2 Prepare slides of the cells. The best method of preparation is by centrifugation of the cells on to the slide using a Cytospin centrifuge (Shandon Scientific Ltd).
3 Fix the cells in anhydrous methanol for 10min at room temperature.
4 Rinse the slides with PBS.
5 Stain the cells with Hoechst 33258 for 10 min at 37°C.
6 Wash the slides in PBS.
7 Examine the slides with a fluorescent microscope fitted with a 420nm filter and X50 water-immersion objective.
8 Mycoplasma are pleomorphic, but are usually observed as small random specks or branched filaments, fluorescent blue, associated with the cytoplasm or at the outer edge of the cell (Figure 3, page 20).

Culture and storage of plasmacytoma cell lines

The plasmacytoma cells are grown in either RPMI 1640 or Dulbecco's Minimum Essential Medium (DMEM) containing 10% FCS and supplemented with pyruvate, glutamine and antibiotic (see page 23). In our laboratory RPMI 1640 is routinely used. The cell viability is very density-dependent and should be maintained at $1 \times 10^5 - 8 \times 10^5$ cells/ml. The cells are usually cultured in 75 cm^2 tissue-culture flasks (50 ml/flask) stood upright at 37°C in a 5% CO_2 atmosphere. Some plasmacytoma cell lines prefer a 7% CO_2 atmosphere. As the cells double in less than 24h, they require subculturing every 2 days. If it is not possible to attend the

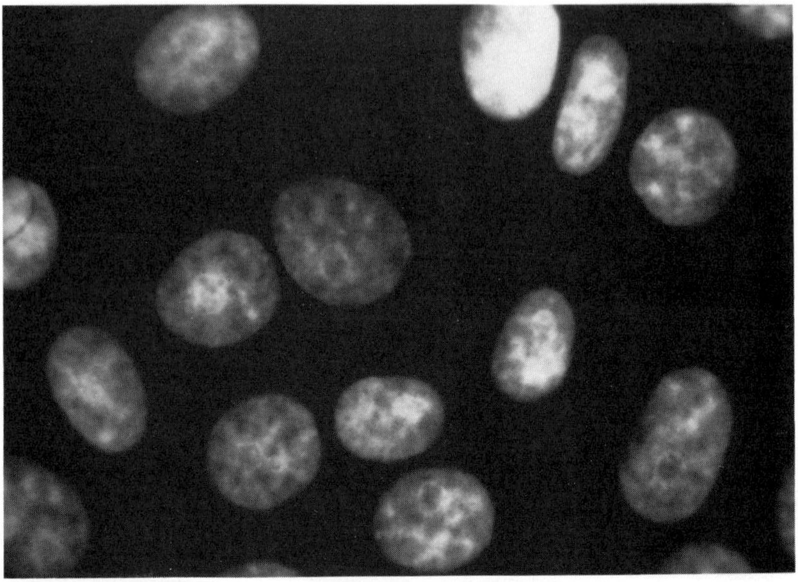

Non-infected control

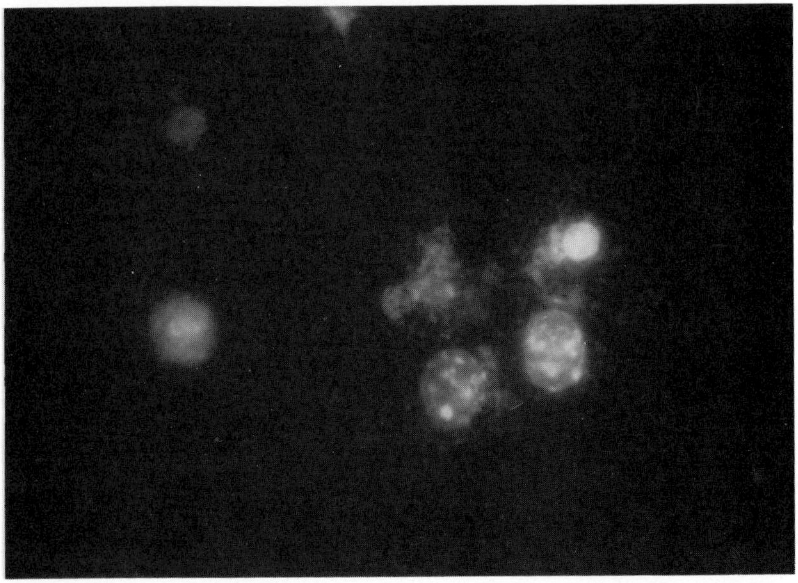

Mycoplasma-infected cells

Figure 3 Detection of mycoplasma with Hoechst 33258

cells over a weekend, then they can be diluted to 5×10^4 cells/ml on Friday afternoon and fed early on Monday morning.

To be safe, it is advisable to grow up to 1l of culture and freeze 30-50 vials, containing 2ml aliquots at 5×10^6 cells/ml in FCS containing 10% dimethyl sulphoxide (DMSO), and store them in liquid nitrogen (see page 58).

Preparing plasmacytoma cells for fusion

From 7 to 10 days prior to the fusion, a vial of frozen cells is placed in sterile water at 37°C to quickly thaw the cells. The vial is swabbed with 70% alcohol and the contents are then rapidly transferred into 20ml of medium containing 10% FCS. The suspension is centrifuged at 150g for 10min, the supernatant is discarded and the cells are resuspended in 10ml of fresh medium with 10% FCS. This cell suspension is cultured in a 25 cm^2 tissue-culture flask and incubated overnight at 37°C. The cells are then diluted to 10^5 cells/ml and cultured as previously described. Three days prior to the fusion, the cells are seeded at 1×10^5 cells/ml into four 75cm^2 flasks (50ml/flask). Alternatively, the cells are seeded at 1×10^5 cells/ml into a spinner culture vessel (Techne Ltd) containing a total of 200ml of medium.

Twenty-four hours prior to fusion, the cells are diluted to $1\text{-}2 \times 10^5$ cells/ml so that they are in the logarithmic phase of growth required for optimal fusion efficiency.

Immune cell fusion partners

The spleen is the major lymphoid organ involved in the generation of the blasting B-lymphocytes suitable for fusion. Being directly in the pathway of the main blood flow, the spleen is also perfused, rapidly, with a substantial proportion of the antigen introduced into the body by intraveneous injection. In most cases, therefore, the spleen is the preferred source of lymphocytes for fusion experiments.

Other lymphoid tissues

There are circumstances in which lymphoid organs other than the spleen may be chosen as the source of immune lymphocytes in fusion experiments—for example, Peyer's patches and mesenteric lymph nodes.

Many of the antigenic challenges to which the body is subjected occur at mucosal surfaces, especially in the upper respiratory, urogenital and gastrointestinal tracts. The predominant isotype generated at these sites of antigenic stimulation is IgA, and in these cases the regional lymph nodes become the centre of antibody generation.

Particulate antigens may be orally administered and the lymphocytes may be recovered from gut-associated lymphoid tissues (GALT) and used in fusions to give hybridomas secreting monoclonal IgA antibodies (Komisar *et al* 1982). However, a substantial proportion of the hybridomas generated by fusing the spleen cells of mice similarly immunized will also secrete antibody of the IgA isotype (Colwell *et al* 1986).

Reagents

Polyethylene glycol

Polyethylene glycol (PEG) causes aggregation of the integral proteins in the cell membrane, resulting in areas of lipid bilayer denuded of protein. Two juxtaposed areas of such denuded membrane will merge, which will allow the fusion of adjacent cells (Galfre *et al* 1979). PEG of molecular weights between 1,000 and 6,000 may be used successfully in fusion experiments, but the size of PEG selected can affect the fusion frequency: PEG of lower molecular weight is incorporated more readily into the cell membrane, but may traverse the membrane, causing cytotoxicity; PEG of higher molecular weight tends to be less cytotoxic.

Additionally, there is considerable variability in the purity and fusing efficiency of batches of PEG. For this reason, once a suitable batch of PEG is identified, either by trial and error or by recommendation, large stocks should be obtained.

A 40% (v/v) PEG solution for fusion is prepared as follows.

1 4g PEG (molecular weight 4000; Merck Ltd) is autoclaved at 15lb for 15min, then allowed to cool to 37°C.
2 6ml of serum-free medium at 37°C is added and mixed thoroughly. The pH is adjusted to approximately 7.2 with 1M sodium hydroxide, using the colour of the medium as a guide.
3 1ml aliquots of the solution are stored at 4°C until required.
4 10% (v/v) DMSO may be added to the PEG solution but is not essential for successful fusions.

Fetal calf serum

The biological properties of batches of FCS vary markedly. All serum should have been screened for mycoplasma and bovine adventitious agents at source. Before purchase, request and test serum samples. An indication of the suitablity of serum for fusion experiments is obtained by cloning the parental plasmacytoma cells (see page 31) in medium containing the sample, ideally in comparison with a previously acceptable serum. Both rate of cell growth and the cloning efficiency are indicators of serum suitability. It is also possible to purchase prescreened batches of FCS—for example Myclone FCS (Gibco Ltd).

Serum for tissue culture should always be heat-inactivated for 30min at 56°C before use, in order to destroy any complement activity.

Media

Two standard media are used routinely in hybridoma culture—RPMI 1640 and DMEM. Both are normally used in laboratories equipped with CO_2 gassed and humidifed incubators. HEPES (15mM) may be included in the medium to provide extra buffering capacity. HEPES-buffered medium is recommended when cells are cultured in gassed CO_2 chambers or other non-humidified CO_2 atmospheres. The HEPES medium also avoids the pH extremes experienced by cells while the plates are microscopically scanned, and reduces the risk of fungal contamination by eliminating the need for high humidity. The media may be purchased as single strength; if ×10 concentrated solution or powder is used, water, glutamine and bicarbonate must be added to give a single-strength solution.

The use of ×10 concentrated RPMI 1640 media is recommended and should be made up by mixing together the following:

10ml ×10 RPMI 1640 (purchased sterile)
86ml sterile, double distilled deionized water
1.1ml 7.5% (w/v) sodium bicarbonate, (filter-sterilised and stored at 4°C)
1.0ml 200mM glutamine, (filter sterilized and stored at −20°C in 1.0 ml aliquots)
1.0ml 10mg/ml sodium pyruvate (filter sterilized and stored at −20°C in 1.0ml aliquots)
40μl 250mg/ml kanamycin (Winthorp Ltd) (alternatively, 100IU/ml penicillin with 10μg/ml gentamycin)

Note
Glutamine, sodium bicarbonate and sodium pyruvate can be bought as sterile solutions. 1.5ml 1M HEPES may be added for use in non-humidified CO_2 atmospheres.

Media supplements

Two media supplements are used in the early stages of hybridoma production—HT and HAT. Both are made up as ×100 stock solutions. (Both HT and HAT sterile stock solutions are available commercially.)

HT×100 stock solution 136mg hypoxanthine and 38.7mg thymidine dissolved in 100ml of double-distilled, deionized water warmed to 70-80°C. Sterilize by filtration through 0.22μm filter. Store at −20°C in 1ml aliquots (enough for 100ml medium).

HAT×100 stock solution Prepare HT stock solution as above adding 1.76mg aminopterin. The aminopterin may not readily dissolve, in which case add a few drops of 1M sodium hydroxide. Sterilize and store as before. Aminopterin is unstable in solution, so HAT medium should be made up fresh.

Feeder layers

Hybridoma cells, particularly newly fused cells, grow very inefficiently in culture at low cell density. So clones of these cells are usually cultured in the presence of feeder cells, which provide certain growth factors and a suitable microenvironment to encourage hybridoma growth. Ideally the feeder cells should not themselves divide, and macrophages or thymocytes are commonly used.

Hybridoma growth factors, which replace feeder cells, have recently become available commercially. However, such factors are currently extremely expensive and do not appear to give reliable results in comparison with feeder cells; they are not recommended.

Macrophages feeder layers Peritoneal macrophages are recommended, as they are easily obtained and are long-lived, adherent and phagocytic, so that the culture is cleaned of dead cells.

Method

1 A mature mouse (older mice give the best yields) is killed by cervical dislocation, soaked in 70% alcohol and pinned to a dissection board.

2 The skin is dissected away from the abdomen and pinned back, exposing the peritoneum.

3 The peritoneal membrane is swabbed in 70% alcohol, and 5ml of cold, serum-free medium is injected, over the liver lobes, into the peritoneal cavity with a 21g × 1in needle.

4 The abdomen is massaged gently for 2-3min. Using a syringe and 21g × 1 inch needle, as much fluid as possible is recovered from the peritoneal cavity; the needle is inserted on the left side of the mouse, just above the leg, so that blockage of the needle with loose fat and tissue is minimized. Care must be taken to avoid puncturing the intestines.

5 The cell suspension is centrifuged at 150g for 10min and then resuspended in 40ml of medium containing 20% FCS (final concentration 2-5 × 10^5 cells/ml).

6 100μl per well of the cell suspension is dispensed in the centre 60 wells of 6, 96-well tissue culture plates. (In a non-humidified CO_2 atmosphere the outer wells can become dehydrated. These outer wells should be filled with sterile distilled water containing ×2 concentration of antibiotics.)

7 When producing feeder layers in 24-well plates for clone expansion, the cells are resuspended in 75ml of medium. This is sufficient for 3 plates at 1ml per well.

8 Macrophage feeder layers should be incubated for 24h before use, to check sterility, and may be used up to 2 weeks after production.

Thymocyte feeder cells Thymocyte feeder cells have the advantage that they die within 7-10 days in culture, so that the hybridoma clones are more easily observed. A thymus from a recently weaned mouse provides approximately 10^8 thymocytes which is enough to fill eight 96-well plates.

Method

1 A mouse (4-8 weeks old) is sacrificed by cervical dislocation, pinned ventral surface uppermost to a dissection board and swabbed in 70% alcohol.

2 The skin is dissected away from the abdomen and thorax, the skin flaps are pinned back and the underlying tissue is swabbed with 70% alcohol.

3 The peritoneum is opened and scissors are used to cut through the ribs which are removed to expose the heart and lungs. Care must be taken not to puncture the heart. The thymus is a cream-coloured bilobed organ, found lying above the heart (see Figure 4).

4 The thymus is removed and immediately placed in 10 ml of medium containing 20% FCS. The thymus, with the medium, is dissociated by pushing the tissue through a sterile tea strainer or steel mesh, placed in a petri dish, with the barrel of a 5ml plastics syringe.

5 The cells are diluted to 85ml in medium with 20% FCS (final concentration 1-2 $\times 10^6$ cells/ml).

6 100μl aliquots are dispensed per well of eight 96-well plates and incubated at 37°C in 5% CO_2. (The outer wells can be filled with distilled water containing $\times 2$ concentration of antibiotics to reduce dehydration in non-humidified CO_2 atmospheres.)

7 Thymocyte feeder cells should be used within 24h of production.

Fusion procedure

Materials

RPMI 1640 medium; serum-free single-strength with glutamine and antibiotic
RPMI 1640 medium containing 20% FCS
PEG solution, at 37°C
HT and HAT $\times 100$ stock solutions
96-well, flat-bottomed tissue culture plates containing feeder cells
50ml sterile plastics centrifuge tubes
25ml sterile plastics conical-bottomed universal tubes
60mm sterile plastics petri dishes
Sterile dissecting instruments
70% alcohol
Sterile tea strainer or stainless steel grids (see page 14)
Sterile 1ml, 5ml and 10ml pipettes
8-channel, variable-volume, 50-200μl automatic pipetter
Sterile plastics 200μl pipette tips; preferably in a 96-tip carrier
Stop-clock
5ml sterile plastics syringe
Dissecting board and pins
CO_2 incubator or tissue culture chambers (Flow Laboratories Ltd) gassed with 5% CO_2 in air

Plastics sandwich boxes containing a vessel with 10-20ml copper sulphate solution (approx 1%)
Laminar flow cabinets; preferably one for mouse work and one for tissue culture only

Preparation of spleen cells

1 Anaesthetize the immunized mouse with diethyl ether vapour.
2 Bleed the mouse by cardiac puncture using a 1ml syringe fitted with 23g × 1in needle. Up to 1ml of blood can be collected, to give serum which is used in the screening assay.
3 Kill the mouse by cervical dislocation, pin onto a dissecting board ventral surface uppermost, and swab the fur thoroughly with 70% alcohol.
4 In a laminar flow cabinet the skin is dissected away from the peritoneum and pinned back (preferably in a laminar-flow cabinet separate from that to be used for tissue culture).
5 The peritoneal membrane is swabbed with 70% alcohol, and with fresh sterile instuments the peritoneal cavity is opened and the membrane pinned back, exposing the spleen on the left side of the mouse (see Figure 4).
6 The spleen is aseptically removed and any fatty or connective tissue is dissected away before the organ is placed in 50ml of serum-free medium.
7 The tea strainer or steel mesh is placed in a petri dish and the spleen and medium are forced through the mesh, by gentle rubbing with the barrel of a 5ml plastics syringe.
8 The cell suspension is placed in a 50ml tube and allowed to stand for 5min so that the large tissue fragments settle out. The upper cell suspension is recovered.
9 A 100μl sample of the cell suspension is removed for counting. In order to lyse the contaminating red cells and allow the spleen cells to be readily distinguished microscopically, 2ml of 2% acetic acid is mixed with the cell sample prior to counting. Approximately 1-2 × 10^8 cells should be obtained from a hyperimmunized mouse.
10 The cell suspension is washed three times in 50ml volumes of serum-free medium by centrifugation at 150g for 10min.

Preparation of plasmacytoma cells

1 Plasmacytoma cells in the logarithmic phase of growth (see page 21) are harvested from the flasks. Some of the cells will be lightly attached to the plastics and need to be resuspended in a vigorous flow of medium from a pipette.
2 Once harvested the cells are counted and washed three times in 50ml volumes of serum-free medium by centrifugation at 150g for 10min.

Fusion technique

1 The washed plasmacytoma and spleen cell suspensions are mixed, to

give a ratio of between 1:1 and 1:4 plasmacytoma:spleen cells, in a 50ml centrifuge tube.

2 The mixed cell suspension is centrifuged at 150g for 10min.

3 All of the supernatant is carefully removed from the cell pellet.

4 The cell pellet is dissociated by gently tapping the outside of the tube.

5 With a sterile 1ml pipette, 1ml of PEG solution is slowly added over a 1min period while stirring with the tip of the pipette. The cell suspension should acquire a granular appearance.

6 The cell suspension is allowed to stand for 1min.

7 45ml of serum-free medium is added, over a 5min period, from a pipette, while stirring with the pipette tip.

8 The cell suspension is centrifuged at 150g for 10min. The supernatant is removed and the cell pellet is disaggregated by gently tapping the tube. The cells are resuspended very gently in 10ml of medium containing 20% FCS and ×2 HT supplement. The small cell clumps, which contain the fused cells, are very fragile and should not be dissociated.

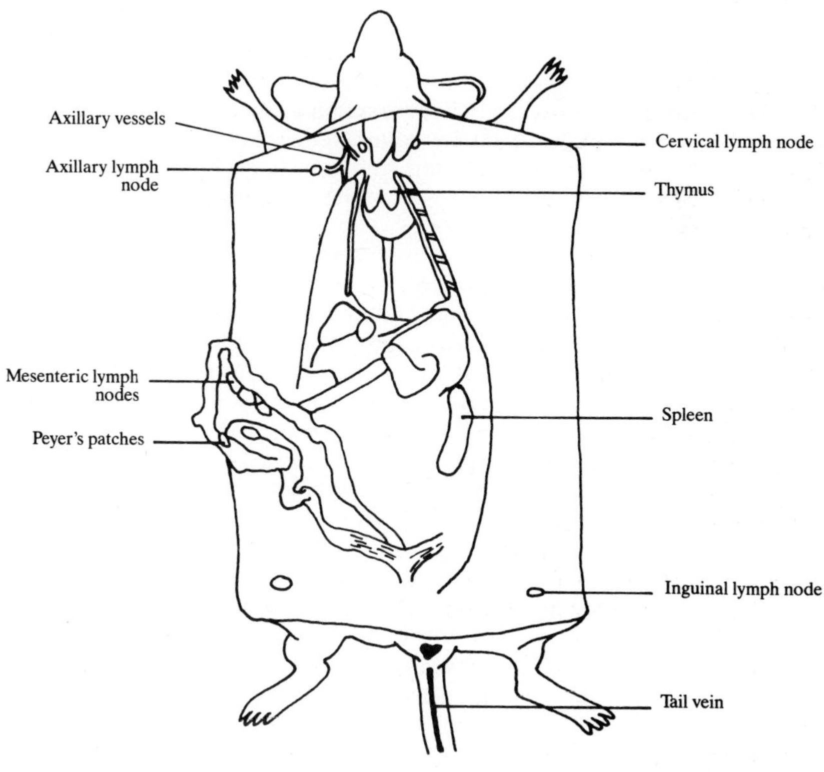

Figure 4 Dissection of mouse for primary and secondary lymphoid tissue

9 This cell suspension is incubated at 37°C for 2h.
10 Following incubation, the cells are centrifuged and gently resuspended as before but in 40-100ml (see note below) medium containing 20% FCS and ×2 HAT.
11 100μl of this cell suspension is pipetted into the centre 60 wells of from 6 to 12, 96-well tissue culture plates (see note below) already containing 100μl of medium and feeder cells (either macrophages or thymocytes).
12 Number the plates (these are the master plates) and incubate them at 37°C in unsealed, humidified plastics sandwich boxes in a 5% CO_2 incubator or in small chambers gassed with 5% CO_2 in air.

Note The volume of medium used to finally resuspend the cells is determined by several factors. If it is preferred that the first growth phase should also be a cloning phase, then the number of potential clones per well is reduced by increasing the volume and, of course, the number of wells seeded. In this case, the cells should be resuspended in 100ml of medium and distributed through 12 plates. On the other hand, some workers consider that multiple clones per well improves hybridoma survival chances. In this case, the number of plates handled can be reduced to as few as 6 (40ml). The multiple-clone philosophy has the additional advantage of reducing the screening load, but means that at least three cloning cycles must be completed to ensure monoclonality. Moreover, it should be remembered that non-Ig-secreting hybridomas have a growth advantage over Ig-secreting clones and may therefore quickly outgrow them.

Caring for hybridomas after fusion

1 The cells are fed every 2-3 days after fusion by removing 100μl of spent medium and gently replacing this with 100μl of fresh medium containing HAT and 20% FCS. This re-feeding can be efficiently done with a multi-channel pipetter. Great care must be taken not to disturb the clones. This early and repeated feeding increases the growth rate and allows earlier screening to take place.
2 After 14 days in culture, the cells are re-fed with medium containing HT supplement instead of HAT. HT should be included for at least a further 7 days.
3 Clonal growth is monitored with an inverted microscope, preferably fitted with a large stage to take plates. Clones containing 10-20 cells should be clearly visible with a ×4 objective by 10 days in culture (Figure 5 a-d).
4 Wells containing clones can be screened for specific antibody production from 10 days onwards (see Chapter 4).
5 Wells containing antibody-positive clones should be cloned as early as possible.
6 Antibody-positive wells are initially designated as the fusion

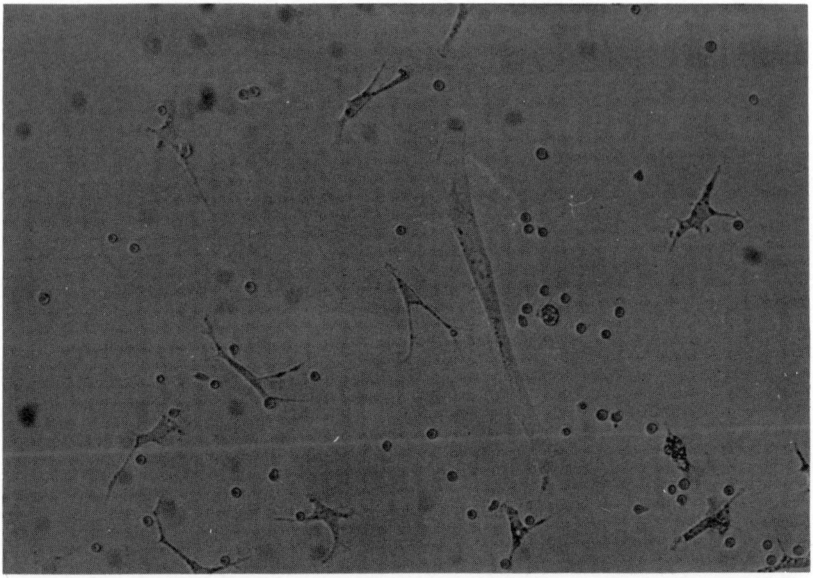

Figure 5a Macrophage feeder layer. Magnification ×4

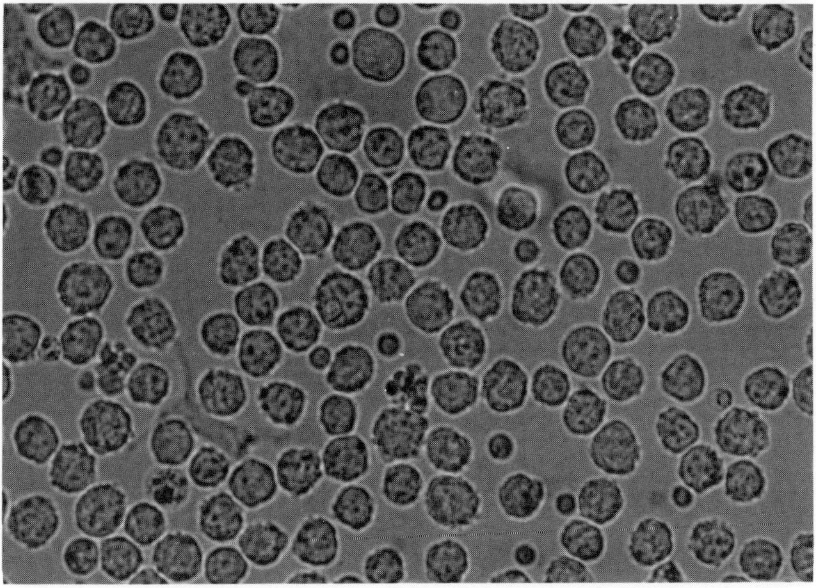

Figure 5b Three-day culture of hybridomas on thymocyte feeder layer. Magnification ×20

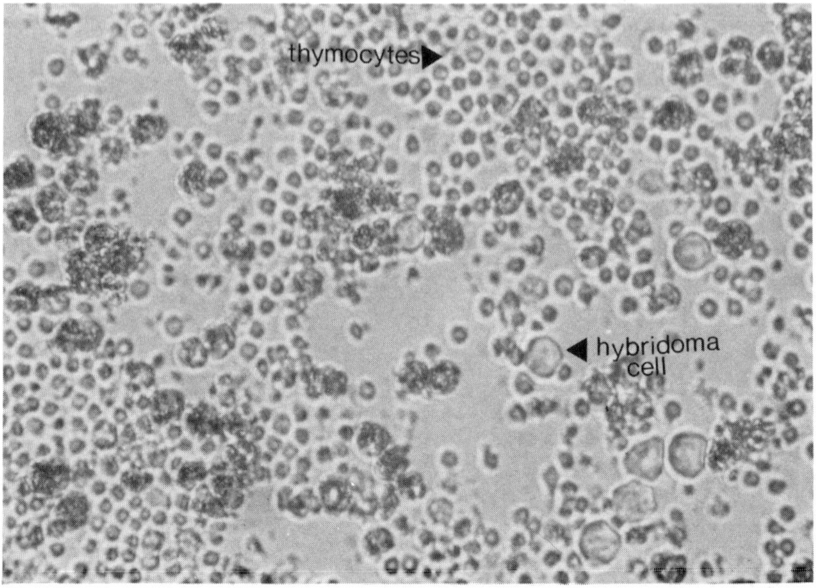

Figure 5c Seven-day culture of hybridomas on thymocyte feeder layer. Magnification ×20

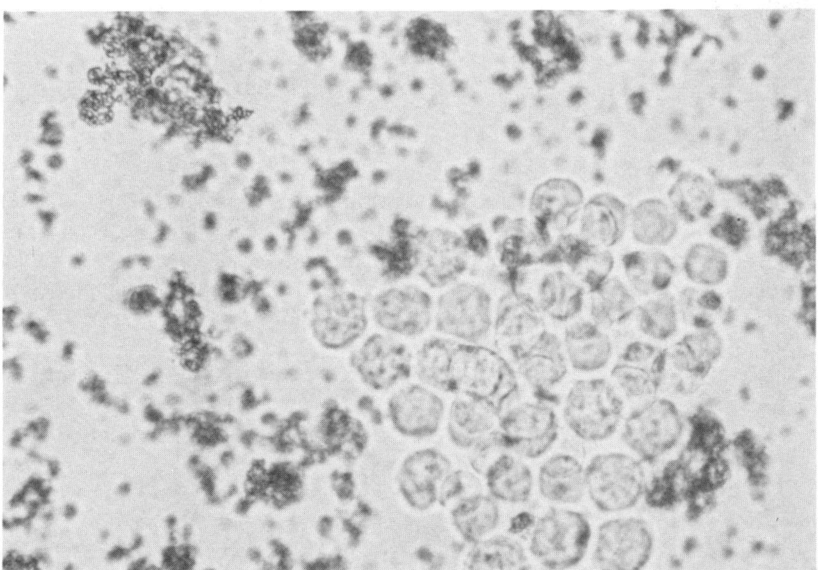

Figure 5d Hybridoma colony ready for expansion on macrophage feeder layer. Magnification ×20

Erratum

Figure 5 The photographs and captions on pages 29 and 30 should be arranged as this sheet

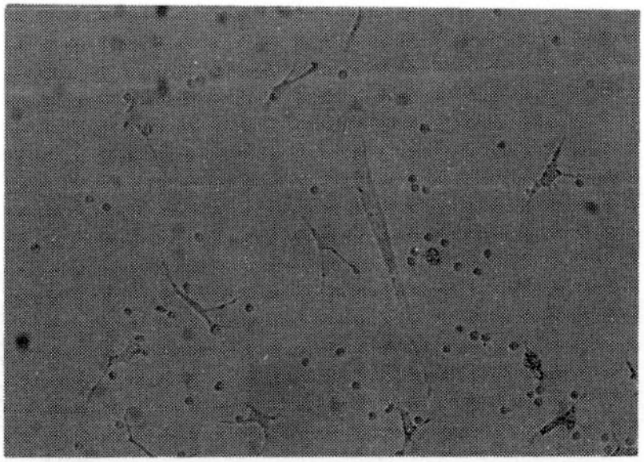

Figure 5a Macrophage feeder layer. Magnification ×4

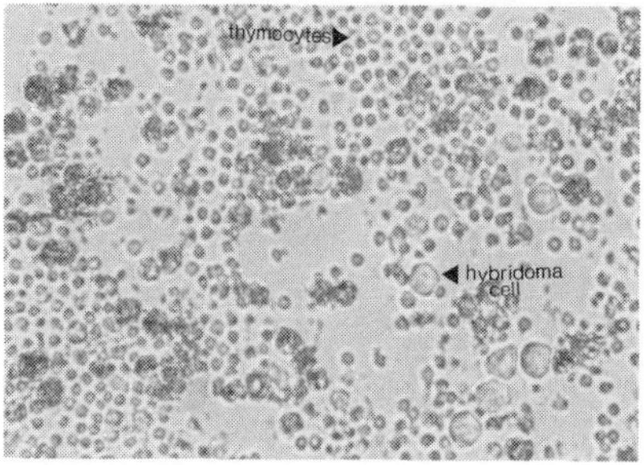

Figure 5b Three-day culture of hybridomas on thymocyte feeder layer. Magnification ×20

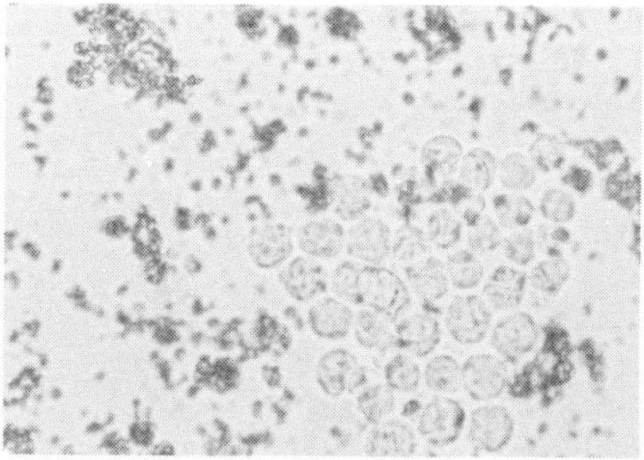

Figure 5c Seven-day culture of hybridomas on thymocyte feeder layer. Magnification ×20

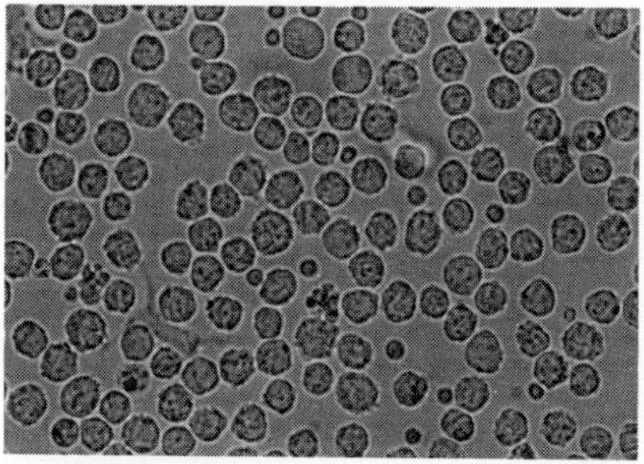

Figure 5d Hybridoma colony ready for expansion on macrophage feeder layer. Magnification ×20

experiment number, followed by the master plate and well numbers—eg 5/3F4. However, we strongly recommend that useful clones be given a final laboratory number, preferably related to the antigen, to avoid future confusion.

Cell cloning

When a particular well has been identified as containing specific antibody-secreting hybridomas, it is important to immediately isolate a single antibody-secreting cell. This cell is then cultured, to produce a clone of identical cells. This ensures that the antibody produced is truly monoclonal and that the cells are sufficiently robust for future use. If multiple clones were grown in the master plates, then monoclonality can only be guaranteed if the cells are cloned at least three times.

Dilution cloning

The simplest method of cloning is stepwise dilution: a low concentration of cells is double diluted through the rows of a 96-well tissue culture plate past the point where each well contains one cell.

Materials

96-well tissue culture plates containing macrophage feeder cells, in RPMI 1640 medium with 20% FCS, in centre 60 wells
Microtitre plate reading mirror (Flow Laboratories Ltd) (optional)
10-200μl variable pipetter and sterile pipette tips

Method 1

Dilution cloning of master plate wells containing only 1-3 discrete clones per well

1 An extra 100μl of medium with 20% FCS is placed in each of the first wells (B2-G2) of the 96-well tissue culture plate containing feeder layer cells.
2 A clone of cells is located in the master plate, either by eye or with the aid of a microtitre-plate reading mirror. These mirrors are easily placed in the laminar-flow cabinet and allow the clones to be viewed from underneath the plate.
3 With a small sterile pipette tip 10μl of cells from the clone is aspirated into the tip.
4 The cells are placed in the first well (B2), which contains the extra 100μl of medium, and thoroughly mixed.
5 100μl of this cell supension is transferred to the next well in the row.
6 This doubling dilution is continued until all wells are used.
7 100μl of medium with 20% FCS is added to all wells, so that each well contains 200μl of medium.
8 The plate is incubated for 7 days and clonal growth is assessed. The wells furthest across the plate should contain only a few clones. These

wells are assayed for specific antibody.
9 Antibody-positive wells with between one and three clones are recloned.
10 In order to ensure monoclonality, it is necessary to repeat this cloning procedure at least twice.
11 After the final cloning the clone should be expanded (see page 44) to provide supernatant for characterization and cells for storage and ascites production.

Method 2
Dilution cloning of master plate wells containing multiple clones

The technique is exactly the same as Method 1, except that all of the cells in the master plate well to be cloned are resuspended in the medium. Aliquots (10-50μl, depending on cell density) are removed and transferred to the top wells of the first three rows (B2-D2) of the dilution plate. The cells are double diluted down the row of wells as before. Each master plate well then occupies one third of a dilution plate. The cells are incubated and screened as before. With this technique to ensure monoclonality this cloning procedure will need to be repeated at least twice more.

Method 3
Like Method 2 except that the cells are counted. Dilutions are made so that approx 10, 5, 3 or 1 cells are added to each well of a 96-well plate.

Soft agar cloning
For the majority of hybridomas, cloning by the dilution method is recommended. But hybridomas secreting antibody which precipitates soluble antigen may be more suitably cloned by the soft agar method.

Materials
Seaplaque agarose (Miles Research Labs)
RPMI 1640 medium
FCS
20mm petri dishes
Dry ice
1 BALB/c mouse

Method
1 0.17g agarose is dissolved in 5ml double distilled water, by autoclaving at 15psi for 15min.
2 75ml of RPMI 1640 medium, warmed to 43°C, is added slowly to the hot agarose solution. Then 20ml of FCS is added, with shaking.
3 4ml aliquots of the agarose solution are quickly poured into 15 petri

dishes and allowed to set for about 30min.

4 A beaker of dry ice, with a little water added, is placed in the laminar-flow cabinet among the plates. The dry ice and plates are left to stand in a closed sterile cabinet with the fans off. This will accelerate the setting, and help to maintain the correct pH, of the agarose.

5 Peripheral blood is collected from the mouse by cardiac puncture, as described on page 26.

6 The blood is diluted immediately, 1:100 in RPMI 1640 medium containing 20% FCS, warmed to 37°C.

7 1.5×10^4 of the hybridoma cells to be cloned are centrifuged and the cell pellet is resuspended in 5ml of the diluted mouse blood solution.

8 10ml of agarose/RPMI 1640/FCS solution (at 37°C) is added to the cell mixture, to give a final volume of 15ml.

9 1ml aliquots of this cell-agarose solution are pipetted onto the solidified agarose in the petri dishes and allowed to set for about 20 min as previously described.

10 The cells are incubated in a humidified box at 37°C in 5% CO_2 for 4-6 days. Discrete clones of cells grow in the upper agarose layer and can be observed with an inverted microscope.

11 After 4-5 days in culture, 0.2ml of sterile soluble antigen is overlaid on the agar surface. (The concentration of antigen required will need to be determined by titration.)

12 Precipitates of antibody-antigen complexes are formed over the antibody-secreting clones over a 7-day incubation period.

13 Positive clones are selected with a sterile, long-formed glass pasteur pipette and the cells are transferred to fresh medium.

14 To ensure monoclonality, this cloning procedure should be repeated at least twice.

15 After the final cloning, the clone should be expanded (see page 44), to provide cells and supernatant.

Chapter 3

Screening

It cannot be over-emphasized that the production of monoclonal antibodies to a particular organism should not be carried out unless a good screening system is already available. The screening system should ideally have the following characteristics.

It should employ the same form of antigen that has been used in the immunization regimen. This means that if the antigen used for the immunization was formaldehyde fixed, for instance, then the antigen used for screening should be treated with formaldehyde.

The screening system used should preferably be of the same type as that proposed for the diagnostic kit. For example, if the monoclonal antibody is to be used for the detection of cell surface antigens by histochemical staining then this technique should be used as the assay system. This may be impossible, however, owing to the restricted availability of valuable purified antigen or the tedious nature of the assay. In these circumstances the screening can be carried out with a crude antigen or using a rapid assay. Subsequently promising hybridomas are screened with a more specific assay once the number of clones has been reduced. It should always be borne in mind that a monoclonal antibody that is detectable in one assay—for example an enzyme-linked immunosorbant assay (ELISA)—may not function in another assay, such as immunofluorescence.

The assay used for screening should be rapid and capable of handling large numbers of samples. On the initial screen several hundred wells will have to be tested, assuming that the fusion has been successful.

This initial screening may have to be repeated two or three times before all positive wells have been selected. Moreover, large numbers of wells may have to be screened at clonal selection.

Unless the monoclonal antibodies are to be used for such procedures as affinity chromatography, the assay should be designed to detect high avidity antibodies. This can be complicated by the rate of antibody production by the cells. Strong positives may be due to either a low concentration of high avidity antibody or a high concentration of antibody of relatively low avidity. The use of short incubation times and mild detergent washes when reacting the monoclonal antibody with the antigen may help to eliminate these anomalies. The sensitivity of the assay will depend on the antigen used in the test. A highly sensitive assay which uses an antigen comprising many specific epitopes will detect low—as well as high—avidity antibodies. Conversely, if the antigen has only a few specific epitopes—for example, antigens on cell surfaces—then a relatively high sensitivity will be required for the signal to be detectable.

Since the initial screen will detect relatively few positive hybridomas from a large number of wells, some consideration should be given to the cost of the assay.

A large number of assays have been developed for the screening of supernatants, and the selection of an appropriate assay will depend on the criteria established above. The methods which follow are those which are in general use, but the list cannot be considered exhaustive.

Histochemical staining

Tissue for paraffin embedding should be fixed in Carnoy's fixative for 20 min before processing. It should be noted that some monoclonal antibodies may not react with tissues fixed by this method. Cryostat sections should be fixed in water-free acetone at 4°C for 10min and air dried. If cryostat sections are to be stored for more than 3 days, they should be fixed in acetone for 10s, wrapped individually in silver foil and stored at −20°C over silica.

Materials

Carnoy's fixative: 60% ethanol
30% chloroform
10% glacial acetic acid

Tris—buffered saline (TBS) pH 7.2: 16.0g NaCl, 1.2g Tris
approx 9ml 1M HCl
made up to 1l

Phenylhydrazine: 0.1% (v/v) in TBS

Walpole buffer: 79 ml 0.1M sodium acetate trihydrate
21ml 0.1N acetic acid

3-Amino-9-ethyl carbazole (AEC): 5mg dissolved in 3ml DMSO. To 25ml Walpole buffer add 10μl 30 volume H_2O_2 then add the AEC in DMSO. Filter through a 0.22μm filter before use. AEC substrate solution must be made up just before use

Glycerol jelly: 100ml distilled water
15g gelatin
100ml glycerol

Dissolve the gelatin in warm distilled water and then add the glycerol.

Mayer's haematoxylin: 1g haematoxylin
50g ammonium or potassium alum
0.2g sodium iodate
1g citric acid
50g chloral hydrate
1 l distilled water

The haematoxylin is dissolved in the distilled water, using gentle heat. The alum is added and shaken with gentle heating until dissolved. The sodium iodate is added, followed by the citric acid and chloral hydrate.

Method

1 A mark is scored around the tissue section on the underside of the slide. This allows easier location of the section during staining.
2 The tissue sections are de-paraffinized and rehydrated in TBS.
3 Endogenous peroxidase activity is quenched with 0.1%(v/v) phenylhydrazine for 1h at 37°C.
4 The slides are washed in TBS three times for 5min.
5 Tissue culture supernatants are applied to the tissue sections, undiluted, and incubated for 1 hour at room temperature in a humid box. Some monoclonal antibodies may require incubation overnight at 4°C.
6 The slides are washed as in 4.
7 Sheep anti-mouse IgG coupled to horseradish peroxidase is added, optimally diluted, to the sections (approx. 1:40-1:100 in TBS) for 30 min at room temperature. The use of $(Fab)_2$ conjugates (Amersham International Ltd) may be of advantage in reducing background staining.
8 The slides are washed as in 4.
9 The AEC substrate solution is applied for 10min at room temperature.
10 The slides are washed in running tap water for 10min.
11 The slides are counterstained in Mayer's haematoxylin for 2min at room temperature.
12 The slides are washed in running tap water for 10min.
13 The slides are mounted in glycerol jelly.

Alternative stain Substrate buffer: 12ml 0.2M Tris
19ml 0.1N hydrochloric acid
19ml distilled water

5mg 3,3′- diaminobenzidine tetrahydrochloride (DAB) is weighed into a glass vial and 10ml Tris—HCl buffer, pH 7.6, is added. Immediately before use, 0.1ml of freshly prepared 1% H_2O_2 (0.1 ml H_2O_2 in 2.9ml distilled water) is added. The substrate solution is filtered through a 0.22μm filter before use. Note that DAB is carcinogenic and gloves should be worn when it is handled.

Enzyme-linked immunosorbent assay for soluble antigens

The enzyme-linked immunosorbent assay (Engvall & Perlman 1971) is probably the most useful assay system for the screening of hybridoma supernatants. The solid phase assay can be adapted for most antigens and enables large numbers of samples to be screened very rapidly. Various parameters have to be established before the assay is used to screen supernatants;

The type of 96-well assay plate has to be considered; in general, two types of plate are used—polystyrene and polyvinyl. No general rule can be given for the selection of the plate. Certain manufacturers produce plates which are treated to enhance the binding of antigens. A selection of plates should be tried with the antigen; the plate giving the highest specific signal to noise ratio should be selected.

The concentration of antigen coated to the wells of the assay plate is critical. Too little antigen will reduce the sensitivity of the assay as not all the antibody in the supernatant will be bound. A high level of antigen will lead to leaching of antigen from the plate, along with bound antibody, during subsequent stages. In general, each well is coated with 100μl of a solution of 1-10μg/ml of antigen (for proteins); the exact level is determined in a chequerboard assay (see page 38) using polyclonal serum from an immunized mouse. Serum taken from an unimmunized mouse acts as a negative control.

Materials

Coupling buffer (pH 9.6): 1.59g sodium carbonate
2.93g sodium bicarbonate
0.2g sodium azide
make up to 1,000ml with H_2O

ELISA wash: 17g sodium chloride
1ml Tween 20
make up to 2,000ml with H_2O

ELISA diluent: 0.6g Tris
1g globulin free bovine serum albumin (BSA)
100ml ELISA wash adjusted to pH 7.4 with glacial acetic acid

Alternatively, 5% (v/v) normal rabbit or sheep serum can be added to ELISA wash (pH adjustment is not required).

Conjugate: Sheep $(Fab)_2$ anti-mouse Ig coupled to horseradish peroxidase (Amersham International Ltd). Stored in 50% (v/v) glycerol at −20°C. Optimal concentration is determined experimentally (approx 1:2,000).

Substrate: Two substrates are commonly used for peroxidase—3,3′,5,5′-

tetramethylbenzidine (TMB) and o-phenylenediamine (OPD). TMB is reported to be non-carcinogenic (Goka & Farthing 1987) and is more sensitive than OPD.

Stock TMB solution: 58mg TMB is dissolved in 10ml DMSO and kept in the dark at room temperature. TMB stock solutions have been shown to be stable for 3 months (Goka & Farthing 1987), and although a yellow tint may develop, this will not affect the sensitivity or background in assays.

Substrate buffer for TMB: 0.1M sodium acetate pH 5.2 (use glacial acetic acid to pH solution)

TMB substrate solution: 99ml sodium acetate buffer
1ml TMB in DMSO
10μl H_2O_2
Substrate solution must be made up just before use.

OPD substrate solution: 10mg o-phenylenediamine dihydrochloride is dissolved in 1ml warm methanol and added to 99ml H_2O containing 10μl H_2O_2. The reaction product is read at 492nm. Note that o-phenylenediamine is carcinogenic.

2M Sulphuric acid: 27ml of concentrated sulphuric acid is added to 500ml of water.

Chequerboard ELISA assay

1 The antigen solution is prepared at approximately 50μg/ml in coupling buffer and 200μl is placed into all 12 wells of row A (A:1-12) of two 96-well plates. 100μl of coupling buffer is placed in all 12 wells of rows B—G (B-G:1-12) and doubling dilutions made of the antigen from row A to row G. 100μl of coupling buffer is placed in all 12 wells of row H (H:1-12). The second plate is treated similarly. The plates are incubated overnight at room temperature covered in cling film.
2 The plates are washed for three 10min periods with 150μl/well ELISA wash. After each wash, the plates are patted dry on a pad of paper towels.
3 1:100 dilutions of the positive and control mouse sera are prepared and 200μl is added to the first well of rows A—H, one plate being used for each serum. 100μl of ELISA diluent is placed in the remaining wells (A-H:2-12) and doubling dilutions are made from well 1 of each row through to well 12.
4 The plates are incubated at 37°C for 1-2h in a humidified box.
5 The plates are washed as in 2.
6 100μl per well of sheep $(Fab)_2$ anti-mouse Ig coupled to horseradish peroxidase, diluted 1:2,000 in ELISA diluent, is added to every well.
7 The plates are incubated for 30min at 37°C in a humidified box.
8 The plates are washed as in 2.
9 100μl per well of TMB substrate solution is added and the plates are

incubated for 10min at room temperature.
10 The reaction is stopped with 50μl 2M sulphuric acid.
11 The plates are read at 450nm with a micro ELISA plate reader (Dynatech Ltd).

The correct antigen concentration should be the lowest dilution which gives the highest antibody titre for the positive serum without giving high backgrounds with the negative serum. Problems with non-specific binding of mouse antibody or conjugate, usually due to insufficient blocking, can be judged from the control wells.

ELISA method

1 The antigen is diluted to the predetermined working strength (see above) in coupling buffer. 100μl is added to all the wells of a micro-ELISA plate.The plate is sealed with cling film and incubated overnight at room temperature.
2 The plates are washed for 3 × 5min with 150μl ELISA wash per well. The plates are patted dry on a pad of paper towels after each wash. 50μl of ELISA diluent is added to each well.
3 50μl of tissue culture supernatants is added to each well. Polyclonal mouse serum diluted 1:100 in ELISA diluent is included as a positive control; a negative control from a well with feeder cells but no hybridoma cells is also included. The plates are incubated at 37°C for 1-2h in a humidified box.
4 The plates are washed as in 2.
5 100μl per well of sheep anti-mouse IgG coupled to horseradish peroxidase, diluted optimally in ELISA diluent, is added and the plates are incubated for 30min at 37°C in a humidified box.
6 The plates are washed as in 2.
7 100μl per well of TMB substrate solution is added and incubated for 10min at room temperature.
8 The reaction is stopped by the addition of 50μl 2M H_2SO_4.
9 The absorbance is read at 450nm.

Alternative conjugate—substrate combinations If alkaline phosphatase enzyme conjugates are used in step 5 the substrate in step 7 should be 1mg/ml p-nitrophenylphosphate dissolved in 9.7% (v/v) diethanolamine buffer pH9.8. The reaction is stopped with 50μl 3N sodium hydroxide. Absorbance is read at 405nm.

If ß-galactosidase conjugates are used, the substrate should be 1mg/ml o-nitrophenyl ß-D-galactopyranoside dissolved in 0.1M sodium borate, 0.08M sodium chloride, 10mM magnesium chloride, pH 8.5. The reaction is stopped with 50μl 1M sodium carbonate. Absorption is read at 420nm.

Incubation times of the above substrates are longer than that of TMB—in the range 30-60min.

ELISA using whole cells as antigen

***Materials** (in addition to soluble antigen assay reagents)*

Poly-L-lysine: 10μg/ml in PBS
Glutaraldehyde: 0.025% v/v in PBS
Phenylhydrazine: 0.1% v/v in PBS
Storage buffer: 100mM glycine containing 1% w/v BSA and 0.01% w/v sodium azide

Method

1 ELISA plates are coated with 100μl/well poly-L-lysine for 1h at 37°C
2 The plates are washed twice quickly, with PBS.
3 The cells are added at approximately 2 × 10^5/well (the exact cell concentration is determined in a chequerboard ELISA) in PBS.
4 The plates are centrifuged at 500g for 8min or left at room temperature for 45min.
5 The plates are washed carefully twice with PBS.
6 100μl/well of glutaraldehyde solution is added and incubated for 15min at room temperature.
7 The plates are washed twice for 5min in PBS.
8 100μl/well of storage buffer is added and the plates are stored at 4°C. Plates may be kept in this buffer until needed.
9 The plates are washed as in 7.
10 100μl/well of phenylhydrazine solution is added for 1h at 37°C
11 The plates are washed as in 7.
12 100μl of tissue culture supernatant per well is added and incubated for 2h at 37°C (ideally, the plate should be placed in a 5% CO_2 environment during incubation).
13 The plates are washed three times for 5min with ELISA wash.
14 100μl/well of sheep $(Fab)_2$ anti-mouse coupled to horseradish peroxidase conjugate is added and incubated for 30min at 37°C.
15 The plates are washed as in 13.
16 100μl/well TMB substrate is added for 20min at room temperature.
17 The reaction is stopped with 50μl 2M sulphuric acid.
18 The plates are read at 450nm or by eye.

Immunoblot assay

The most convenient form of immunoblot assay used in the initial screening of hybridoma supernatants is the dot immunobinding assay.

The antigen of choice is applied directly to either nitrocellulose or positively charged nylon membranes. The mass of antigen applied to the membrane must be determined by a chequerboard assay; the range is normally 1-0.05μg. For convenience the antigen is applied to squares of a grid ruled on the membrane. Polyclonal mouse serum diluted from 1:50 to 1:1000 in two-fold dilutions is used as a positive control; normal mouse serum diluted 1:50 is used as a negative control.

Materials
TBS, pH 7.5: 4.84g Tris
58.48g sodium chloride
made up to 2 litre with distilled water
Adjusted to pH 7.5 with hydrochloric acid
Tween/TBS (TTBS): 0.5ml Tween 20
2,000ml TBS
Blocking buffer: 3% gelatin in TTBS dissolved at 37°C
Incubation buffer: 1% gelatin in TTBS dissolved at 37°C
Substrate: 60mg 4-chloro-1-naphthol dissolved in 20ml ice cold methanol. Add 100ml TBS containing 60μl H_2O_2 just before use.

Method
1 With a Hamilton syringe pipette, 1μl of antigen, at the concentration determined by a chequerboard assay, is placed onto the centre of each square ruled on the membrane sheet. The sheet is allowed to dry.
2 The membrane is incubated in blocking buffer for 1h at room temperature on an orbital mixer.
3 1-5μl of tissue culture supernatant is added to each square and incubated for 5min.
4 The membrane is washed for 30min, with five changes, in TTBS.
5 The membrane is incubated with the secondary antibody, sheep anti-mouse IgG coupled to horseradish peroxidase, diluted optimally (1:1,000-1:3,000) in incubation buffer for 1h at room temperature. The optimal dilution must be determined empirically but higher dilutions will lead to lower backgrounds.
6 The membrane is washed as in 4.
7 The membrane is washed in TBS for 5min.
8 The membrane is incubated with substrate solution at room temperature until the blue colour is sufficiently developed.
9 The membrane is washed in distilled water and dried immediately in blotting paper.

If it is found necessary to incubate the monoclonal antibody for a longer period of time, an aluminium template can be constructed (Smith *et al* 1984) or an inverted microtitre plate can be used as described by Bennet & Yeoman (1983). Such apparatus can also be purchased.

Immunofluorescence

The fluorescent antibody test can be used on viable as well as fixed cells. Because of the relatively large molecular size of the antibody probe, only surface antigens will be labelled on viable cells.

Intracellular antigens can be detected by the use of cells fixed in acetone or methanol, which disrupt the cell membrane. Specimens for examination can range from whole bacteria smeared onto a glass slide and fixed, to live or fixed virus-infected cells. The usual fixative is ice-cold acetone, but other fixatives may produce better results for particular applications. Staining of viable cells will detect antibodies which react to the native protein and therefore may not detect antibodies reactive to determinants whose conformation has been altered by fixation, and *vice versa*.

Two methods, one for viable virus-infected cells and the other for fixed bacteria, are given below, but the optimal conditions for each application must be established experimentally.

Immunofluorescence of fixed bacterial smears

1 Either bacterial suspensions are smeared onto a glass slide or a drop is placed in the wells of a Multitest slide (Flow laboratories Ltd) and air dried.

2 The slides are fixed for 5min in ice-cold acetone. Fixed slides may be stored for up to 8 weeks at −20°C, wrapped individually in aluminium foil.

3 The stored slides are allowed to equilibrate to room temperature.

4 The monoclonal antibody tissue culture supernatants are added and incubated in a 5% CO_2 atmosphere (HEPES buffer may be added to give a final concentration of 0.01M if a CO_2 incubator is not available) for 30min at 37°C in a humidified chamber.

5 The slides are washed three times in PBS for 2min.

6 Rabbit anti-mouse IgG coupled to fluorescein isothiocyanate (FITC) diluted optimally (approx. 1:40-1:80) in PBS is added to each well for 30min at 37°C in a humidified chamber.

7 The slides are washed as in 5.

8 The slides are mounted with PBS and nail varnish is used to seal the coverslip.

9 The slides are examined at 492nm by epi-illumination with an ultraviolet microscope.

Immunofluorescence of live cells

1 75μl of cells (5×10^5/ml) is added to each well of a tissue culture chamber/slide (Miles Scientific) in a growth medium suitable for the cell line. The slide is incubated overnight at 37°C in 5% CO_2.

2 The slides are washed three times with PBS for 2min by gentle aspiration, care being taken to ensure that the slides do not dry at any stage.

3 The cells are infected with virus such that 10-20 plaques per well are produced or 50% of the cells express viral antigens within a 48h period.

4 The slides are washed as in 2.

5 50μl of hybridoma supernatants is added to each well and incubated for 30min at 37°C in 5% CO_2 in a humidified chamber.

6 The slides are washed as in 2.

7 50μl rabbit anti-mouse IgG-FITC conjugate, diluted optimally in maintenance medium (approx. 1:40), is added and incubated at 37°C for

30min in 5% CO_2 in a humidified chamber.
8 The slides are washed as in 2.
9 The slides are mounted in maintenance medium and the coverslip secured with nail varnish.
10 The slides are examined at 492nm with a an ultraviolet microscope using epi-illumination.

Radioimmunoassay

Solid phase radioimmunoassays, using a radiolabelled second antibody instead of an enzyme-linked antibody, are infrequently used owing to the relatively high cost of reagents and equipment and the need for stringent safety conditions in the handling of radioisotopes. Detection in these systems is usually by cutting out individual wells (essential for ß-emitting isotopes) or, if ^{125}I is used, by autoradiography of the plate.

Soluble phase radioimmunoassays require large amounts of radiolabelled antigen but should be considered if the final diagnostic use of the monoclonal antibody is to be in such an assay system. Antigen bound to the antibody can be immunoprecipitated by the use of a second antibody (anti-mouse IgG) coupled to Sepharose 4B beads. These beads are then washed free of unbound radiolabelled antigen before counting.

Chapter 4

Large-scale production, purification and storage of monoclonal antibodies and hybridoma cells

The end-product of all the selection and cloning procedures described in Chapters 2 and 3 will, it is hoped, be a monoclonal hybridoma culture growing in a 0.2ml culture well. From this single well enough cells must be grown for storage in liquid nitrogen. It is advisable that at least six ampoules, divided between two liquid nitrogen freezers, be stored. After the security of the cell line has been assured, the clone can be further expanded for bulk antibody production. Large (milligram) quantities of antibody may be achieved either as an ascitic tumour in the peritoneal cavity of an syngeneic mouse or grown *in vitro* in large culture vessels.

Early stages of clonal expansion

The expansion of the clone from a small well to larger culture vessels can be difficult. The rate of cell growth can vary significantly between clones; the cell density, also, is a critical factor. Fortunately, there are a range of plastics vessels which permit a gentle stepwise scale-up of the cultures. The normal route is:

0.2ml well >1ml well >25cm^2 flask (up to 15ml)
>75cm^2 flask (up to 50ml)

At all expansion steps it is important that the cells reach a density of between 5×10^5 and 1×10^6 cells/ml before transfer to the next vessels. During the initial stages of this progression, especially at the lower end of this cell density, feeder cells may be required to ensure that the cells survive the transfer. Once in 25cm^2 flasks, the serum concentration can be reduced from 20 to 10%. Although the main aim is to expand the culture, it is very important to ensure that the cell density does not increase above 1×10^6 cell/ml. Above this cell density the cells are in a distressed state caused by the limiting availability of nutrients in the medium.

Ascites

The hybridomas are derived from plasmacytoma and spleen cells of an inbred strain of mouse (see page 18) and are therefore immunologically compatible with other mice of the same strain. Moreover, the parental plasmacytomas were originally induced in the peritoneal cavity of mice treated with staphylococcal adjuvants (Potter & Robertson 1960). By injecting the hybridoma cells into the peritoneal cavity of mice pretreated with pristane or IFA, rapid cell proliferation occurs, generating an ascitic fluid containing large quantities of cells and high concentrations of antibody (1-10mg/ml).

Materials

Pristane (2,6,10,14-tetramethylpentadecane, Aldridge Chemical Co)
20mM EDTA containing 1% (w/v) sodium azide

Method

1 The mouse is injected ip with 0.5ml of pristane with a 23g × 1in needle.
2 These mice can be used between 4 and 60 days after injection.
3 For each pretreated mouse, approximately 20ml of cell suspension is harvested and centrifuged at 150g for 10min. The cells (approximately 1×10^7 cells) are resuspended in 0.5ml of PBS and injected ip into the mouse with a 21g × 1in needle.
4 Within 7-10 days after injection, the peritoneum of the mouse will become distended and the ascitic fluid can be drained.
5 The mouse is anaesthetized with diethyl ether vapour and a 21g × 1in needle attached to a 5ml syringe is inserted into the peritoneal cavity through the left flank of the mouse. The mouse is gently tilted so that the liquid drains to the site of the needle, and the excess fluid is slowly removed.
6 A 10% (v/v) of EDTA-sodium azide solution can be added to the ascitic fluid, to prevent clot formation.
7 The ascitic fluid is centrifuged for 10min at 200g and the supernatant is recovered and stored.
8 Cells recovered from ascitic fluid may be re-injected into another mouse. This should not be repeated too often as the cells may stop producing antibody.

Note Mice treated in this way rapidly deteriorate in health. The ascitic fluid must be drained regularly (every 1-3 days). Frequently solid tumours are produced in addition to or instead of ascites. Such solid tumour formation tends to be a property of the clone. Mice should be sacrificed as soon as signs of distress are observed. The volume of ascitic fluid obtainable from each mouse will depend largely on the biological properties of the clone, but between 15 and 30ml can be expected.

Alternative procedure

IFA can also be used to pretreat the mice for ascitic fluid production.

1 The mouse is injected with 0.5ml of IFA as above.
2 1×10^6 cells are injected 24-48h later. The ascitic fluid gathered from IFA-primed mice tends to be milky, but by leaving the primed mice for 48h before introducing the cells, this problem is greatly reduced.
3 Ascitic fluid is ready for draining within 4-6 days.

Although this method successfully produces ascitic fluid with fewer cells and in less time than the pristane method, in our experience some monoclonal antibodies lose significant activity in the process.

Bulk antibody production *in vitro*

Although high concentrations of antibody are obtainable from ascitic fluid there are circumstances in which such preparations are inappropriate—for example, the production of monoclonal antibodies for therapeutic use. The alternative is to grow the hybridomas in larger cultures and to harvest the antibody from culture supernatant. If large-scale culture is the preferred procedure, the concentration of serum used in the culture medium may preclude the efficient harvesting of the antibody.

Where less than 100mg of monoclonal antibody is required, a few litres of end-culture supernatant may suffice. End-culture supernatant is that obtained after the cells are grown to the maximum cell density, and includes antibody secreted from viable cells as well as antibody released from lysed cells. Antibody concentrations in such supernatants tend to reach between 10 and 20μg/ml. For this purpose spinner culture vessels (Techne Ltd) can be used. A 1000ml spinner culture flask is seeded with an initial culture volume of 200ml containing 1×10^5 cells/ml.

By adapting the hybridomas to grow in a serum substitute (see next section), cells can be maintained in a serum protein concentration of 1% or less. End-culture supernatants from these cultures may then be concentrated (up to thirty fold) by ultrafiltration (Amicon Ltd) prior to further purification.

It is not within the remit of this book to discuss at length the bulk culture of hybridomas. However, should gram quantities of antibody be required, there are a number of culture systems commercially available. The two main systems currently in use are air lift fermentation (Celltech, UK) and cellular encapsulation (Damon Biotech, USA).

Adaptation of hybridoma cell lines to serum-free medium

For most applications, monoclonal antibody supernatants, prepared in bulk by ascites production *in vivo* or large scale culture in medium containing less than 5% (v/v) FCS, are acceptable. It may be advantageous, particularly when material for therapeutic use is required, to adapt the cells to a defined medium, free of serum. Several of these media are now commercially available—for example, Excell 300 (JR Scientific). Under these conditions hybridomas continue to secrete antibody at a level equivalent to that detected under normal growth conditions. The adaptation procedure requires an intermediate step to reduce the protein concentration in the medium from 10% FCS to 5% of a serum replacement supplement.

Method

1 Hybridoma cells are grown in medium containing 10% FCS to a concentration of 4-8 $\times 10^5$ cells/ml in a 25cm^2 plastics flask.

2 The cells are centrifuged at 150g for 10min and the cell pellet resuspended in 10ml of medium containg 5% controlled process serum replacement type 1 (CPSR-1, Sigma Chemicals Ltd).

3 The cells are then sub-cultured, maintaining a cell concentration of between 1 and 5 $\times 10^5$ cells/ml, for 1-2 weeks in this medium.

4 The cells are grown to a cell concentration of 4-8 $\times 10^5$ cells/ml in a 25cm^2 plastics flask.

5 The cell suspension is centrifuged at 150g for 5-10min and the cell pellet resuspended in 10ml of Excell 300 growth medium (see note 6 below).

6 The cells are sub-cultured, as before, for at least 1 month to ensure that the adaptation is stable.

Notes

1 CPSR-1 contains 35-45mg/ml protein processed from bovine blood; so at 5% (v/v) 1.75-2.25mg/ml of protein is present in the medium.

2 Hybridomas tend to grow more slowly and reach a lower maximum cell density in static culture with Excell 300—for example, 8 $\times 10^5$ cell/ml, compared with 1.2 $\times 10^6$ cells/ml in medium containing 10% FCS.

3 In general, hybridomas adapt poorly to CSPR-1 in the first 1-3 months of culture. However, once established, after the third cloning, very few hybridoma cell lines fail to adapt with the above procedure.

4 On changing to CPSR-1 the majority of cells may die but some cells will survive to grow.

5 Supernatants should be tested at regular intervals during the adaptation process to ensure continued antibody secretion.

6 The address of JR Scientific is 1 Harter Ave, Suite B, PO Box 1862 Woodland, Ca 95695, USA.

Purification of monoclonal antibody from ascitic fluid

Protein-A-Sepharose chromatography

Protein-A is expressed on the surface of *Staphylococcus aureus* and has specific binding sites for the Fc portion of IgG. The avidity of this binding varies between IgG subclasses. By changing the pH of the elution buffer, murine IgG_1, IgG_{2a} and IgG_{2b}/IgG_3 may be sequentially collected. Protein-A is available commercially, either purified from *S aureus* or as a recombinant product (Porton Products Ltd). The technique for subclassing the antibodies is given in Chapter 5.

Materials
2M Tris-HCl buffer, pH 8.0
0.1M Tris-HCl buffer, pH 8.0
0.1M Citrate buffer, pH 6.0
0.1M Citrate buffer, pH 4.5
0.1M Citrate buffer, pH 3.5
Protein-A-Sepharose column: 1.25g freeze dried

Protein-A-Sepharose (Sigma Chemicals Ltd) is swollen in 0.1M Tris-HCl buffer, pH 8.0 to produce a 5ml bed volume column. The column flow rate should be no more than 1ml/min. All procedures should be carried out at 4°C.

Method
1 The Protein-A-Sepharose column is equilibrated with 0.1M Tris-HCl, pH 8.0.
2 The pH of the ascitic fluid is adjusted to pH 8.0 by the addition of 50μl/ml 2M Tris buffer. In order to remove any particulate material, the ascitic fluid is either passed through a precolumn of Sepharose 4B or through a 0.45μm filter. (Occasionally ascitic fluid will require either centrifugation or prefiltration through a 0.8μm filter prior to this.)
3 The ascitic fluid is loaded onto the column. As this is an affinity column, the volume of sample applied to the column is irrelevant. Commercially available Protein-A-Sepharose will bind 10-15mg of immunoglobulin per millilitre. In practice up to 10ml of ascitic fluid is usually run through a 5ml column.
4 The column is eluted with 0.1M Tris-HCl, pH 8.0, until the absorbance has returned to the baseline. The column eluant is retained, and will contain any IgM or IgA present.
5 The column is eluted with 0.1M citrate buffer, pH 6.0, and 1ml fractions are collected into tubes containing 50μl of 2M Tris. This eluant will contain any IgG_1 present in the ascitic fluid.
6 The column is eluted with 0.1M citrate buffer, pH 4.5, and the fractions collected as before. This eluant will contain any IgG_{2a} present.
7 The column is eluted with 0.1M citrate buffer pH 3.5 and the fractions are collected as before. This eluant will contain any IgG_{2b} or IgG_3 present.
8 The column is washed with 20ml of 0.1M Tris, pH 8.0.
9 The protein-containing fractions of the individual peaks are pooled and dialysed against PBS or Tris-HCl buffer, pH 7.2.

Notes

1 Before changing the buffer check that the absorbance of the eluant has returned to the baseline.

2 Always follow the method through to the pH 3.5 buffer wash, to ensure that the Protein-A is free of any bound protein which may decrease the binding capacity of the column in subsequent runs.

3 Always leave the column equilbrated at pH 8.0, to minimize the gradual decrease in the Ig binding capacity.

Figure 6 represents the elution of the IgG_{2a} monoclonal antibody OKT3 by this method. The ascitic fluid used in this instance contained high concentrations of serum. The elution profile shows peaks consistent with normal serum immunoglobulin but with a greatly enhanced IgG_{2a} component.

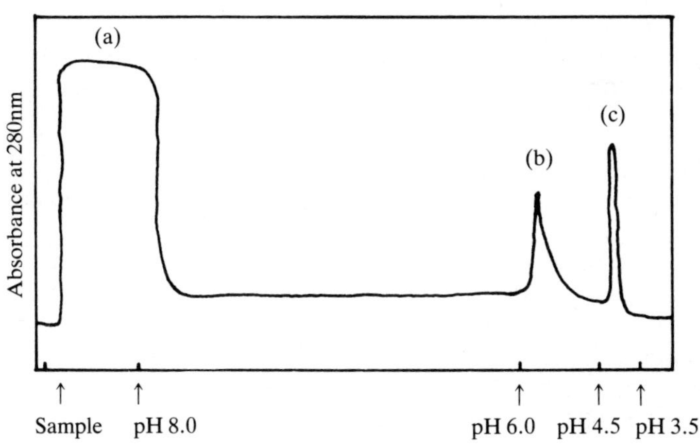

Figure 6 Protein-A – Sepharose column chromatography: purification of IgG_{2a} monoclonal antibody (OKT3) from ascitic fluid (5ml) on a 2ml Protein-A – Sepharose column at 4°C. Peak (a): Application of sample to the columnn at pH 8.0. Peak (b): On changing to pH 6.0, the normal IgG_1, peak was eluted. Peak (c): On changing to pH 4.5, the OKT3 antibody was eluted (yield 15mg)

Note This ascitic fluid contained a higher than normal level of serum, which resulted in the substantial level of IgG_1, collected

Ion-exchange chromatography

Protein-A-Sepharose purification is excellent for all the murine IgG subclasses except IgG_1. Immunoglobulin molecules of the murine IgG_1 subclass bind weakly to Protein-A, which is why they are eluted by a relatively small shift in the pH of the elution buffer. DEAE Tris-acryl M ion-exchange chromatography is a clean and dependable method of purifying IgG_1 monoclonal antibodies. The immunoglobulin is first partially purified by ammonium sulphate precipitation to remove the albumin and then dialysed into low-molarity phosphate buffer prior to loading on the column.

Materials for ammonium sulphate precipitation

Saturated ammonium sulphate solution: 100g ammonium sulphate is added to 100ml distilled water and stirred at room temperature for 24h Before use, the solution is filtered through Whatman No 1 filter paper to remove any undissolved ammonium sulphate

Method

1 The ascitic fluid is centrifuged at 3,000g for 10min.
2 The centrifuged supernatant is filtered through a 0.45μm filter.
3 The filtrate is slowly stirred while the saturated ammonium sulphate solution is gradually added until a 45% (v/v) ammonium sulphate solution is achieved. A precipitate will be produced. The suspension is stirred for a further 20min and then centrifuged at 3,000g for 20min.
4 The precipitate is dissolved in a small amount of PBS.
5 The precipitation process is repeated twice.
6 Finally the precipitate is dissolved in PBS and then dialysed against 5mM phosphate buffer for immediate use or 0.1M Tris-HC1 buffer, pH 7.4, for short term storage.

Materials for DEAE Tris-acryl M ion exchange chromatography

0.1M phosphate buffer: 14.2g/l disodium hydrogen phosphate adjusted to pH 8.0 with EDTA.
5mM phosphate buffer: 0.1M phosphate buffer diluted 1:20
PBS: 7g/l sodium chloride
3.4g/l disodium hydrogen phosphate
0.79g/l potassium dihydrogen phosphate
pH 7.2-7.4
DEAE Tris-acryl M column: 100ml volume column packed with DEAE Tris-acryl M pre-swollen gel (LKB Ltd) equilibrated in 5mM phosphate buffer. The flow through rate should be 50ml/h. The column is run at room temperature. A linear elution gradient will need to be generated, starting with 5mM phosphate buffer and ending with 0.1M phosphate buffer, over a 1000ml volume. Such gradients can be produced by use of three-way pumps or commercial gradient formers.

Method

1 The column is pre-equilibrated overnight with at least 2000ml of 5mM phosphate buffer.

2 The sample of up to 50ml in 5mM phosphate buffer is loaded onto the column. The column capacity should be approximately 100mg globular protein per millilitre of gel.

3 Once the sample is on the column, a gradient of 5mM to 100mM phosphate buffer is used to elute the sample. The total volume of the gradient is 1000ml.

4 10ml fractions are collected.

Note In order to avoid bacterial overgrowth in the low-molarity phosphate buffer, the collection of the fractions should be completed within 48h of equilibrating the column.

Figure 7 shows the elution profile of an IgG_1 monoclonal antibody (OKT6) on a DEAE Tris-acyl M ion exchange column.

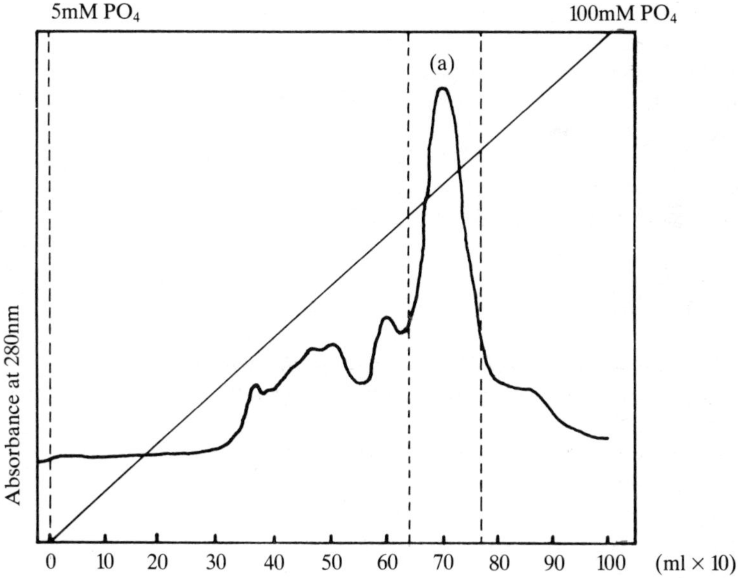

Figure 7 DEAE Trisacryl M ion-exchange gradient column chromatography: purification of IgG_1 monoclonal antibody (OKT6) from ammonium sulphate precipitated ascitic fluid (50ml), on a DEAE Trisacryl M ion-exchange gel. The antibody was eluted by an increasing gradient of phosphate ions from 5mM to 100mM. Peak (a) contained 70mg of OKT6 antibody

Purification of IgM monoclonal antibodies by gel filtration

The purification of monoclonal antibodies of the IgM isotype presents a particular problem. Such antibodies have a tendency to aggregate in solution and rapidly lose activity. The recommended method involves ammonium sulphate precipitation, followed by Ultrogel AcA 22 gel filtration chromatography.

Materials for ammonium persulphate precipitation

Saturated ammonium sulphate solution (see page 50)
0.2M Tris-HC1 buffer, pH 8.0

Method

1 The IgM is precipitated from the ascitic fluid as described on page 50, except that the final concentration of ammonium sulphate should be 55% (v/v).
2 The precipitated immunoglobulin is redissolved in and dialysed against 0.2M Tris-HC1 buffer.

Materials for Ultrogel AcA 22 column chromatography

Ultrogel AcA 22: pre-swollen gel beads (LKB Ltd). A 200ml bed volume column is used with a flow rate of 20ml/h.
0.2M Tris buffer: 0.2M Tris containing 0.1M sodium chloride, adjusted to pH 8.0 with hydrochloric acid.

Method

1 The column is equilibrated in 0.2M Tris buffer.
2 A sample of 10ml, or less, of the ammonium sulphate precipitated IgM (up to 50mg protein) is loaded onto the column. The sample volume should be less than one-twentieth of the column bed volume.

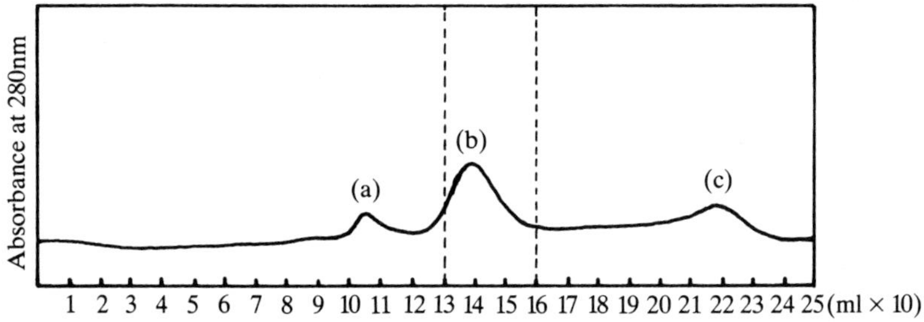

Figure 8 Ultrogel AcA 22 molecular size gel chromatography: purification of IgM for ammonium sulphate precipitated ascitic fluid (20ml) on a Ultrogel AcA 22. Peak (a) IgM complexes. Peak (b) IgM fraction (yield 10mg). Peak (c) residual albumin

3 The column is eluted with 0.2M Tris buffer and 10ml fractions are collected.

Figure 8 represents a typical elution profile of a monoclonal IgM antibody purified on a AcA 22 column of 200ml bed volume. A proportion of the IgM elutes as complexes before the main IgM peak.

Fast protein liquid chromatography

For purification of relatively small volumes of ascites, ion exchange chromatography using fast protein liquid chromatography (FPLC, Pharmacia Ltd) is both quick and efficient. The choice of anionic or cationic exchanger is dependent on the properties of the individual monoclonal antibody and must be determined by trial and error. Monoclonal antibodies of the same subclass may require different types of ion exchange column for purification, but all IgG subclasses may be purified using either MONO Q (anionic exchanger) or MONO S (cationic exchanger) columns.

Materials

For MONO S (column size HR 5/5)
Buffer A (equilibration buffer) 50mM sodium acetate, pH 5.5
Buffer B (elution buffer) 50mM sodium acetate, pH 5.5, containing 0.5M sodium chloride

For MONO Q (column size HR 5/5)
Buffer A (equilibration buffer) 0.005M Tris-HC1, pH 8.0
Buffer B (elution buffer) 0.2M Tris-HC1, pH 8.0, containing 0.2M sodium chloride

Method

1 It is assumed that the reader is familiar with the technique of FPLC.
2 The column is equilibrated with buffer A.
3 One millilitre of ascites is diluted to 10ml with buffer A.
4 The FPLC is programmed with the following method.

Volume	Instruction	Value
0.0	CONC % B	0.0
0.0	ML/MIN	1.0
0.0	CM/ML	0.5
0.0	VALVE.POS	1.2
0.0	CLEAR DATA	
0.0	MONITOR	1
0.0	LEVEL %	5.0
0.0	ML/MARK	1.0
0.0	INTEGRATE	1
0.0	PORT SET	6.1

9.9	VALVE POS	1.1
13.9	CONC % B	0.0
35.0	CONC % B	100
35.0	INTEGRATE	0
35.0	PORT.SET	6.0
39.0	CONC % B	100
43.0	CONC % B	0

5 The diluted ascites is applied to the column with a 10ml superloop.

6 The method is run and 1ml fractions collected.

7 The monoclonal antibody fractions are illustrated in Figures 9 and 10 (overpage). The precise fractions and purity should be checked with a suitable detection system or the Paragon system (see below) using 5μl of the undiluted fractions.

Monitoring the purification of monoclonals from ascitic fluid

It is essential to monitor the antibody concentration throughout the process of purification. Monitoring antibody activity at regular intervals, by a suitable screening procedure, is necessary when the purification techniques may denature the monoclonal protein or when the preparation is contaminated with high concentrations of other proteins, especially immunoglobulins. Such is the case with supernatants concentrated from bulk cultures, unless the cells have been adapted for growth in medium supplemented with little or no FCS. However, the screening procedures are relatively time consuming and give no indication of the degree of contamination with other proteins.

At high concentrations (above 1mg/ml), the monoclonal antibody can be seen as a discrete band, distinct from other contaminating proteins, after gel electrophoresis. Several electrophoretic systems are commercially available, based on thin agarose gels supported with plastics films. Results come within 90min with such systems and allow rapid monitoring of downstream processes.

Figure 11 (page 57) illustrates the results of the purification of several monoclonal antibodies from ascitic fluids, run on a Paragon Electrophoresis System (Beckman Ltd).

Storage of monoclonal antibodies

It is important to remember that storage conditions under which the activity of polyclonal antibodies will survive can easily destroy the activity of monoclonal antibodies. Polyclonals have a range of physical properties, so only a proportion of the antibodies will be denatured by mistreatment; whereas all of the molecules of a monoclonal antibody have the same properties.

Repeated freezing and thawing causes immunoglobulin aggregation, with concomitant loss of antibody activity. It is, therefore, essential to

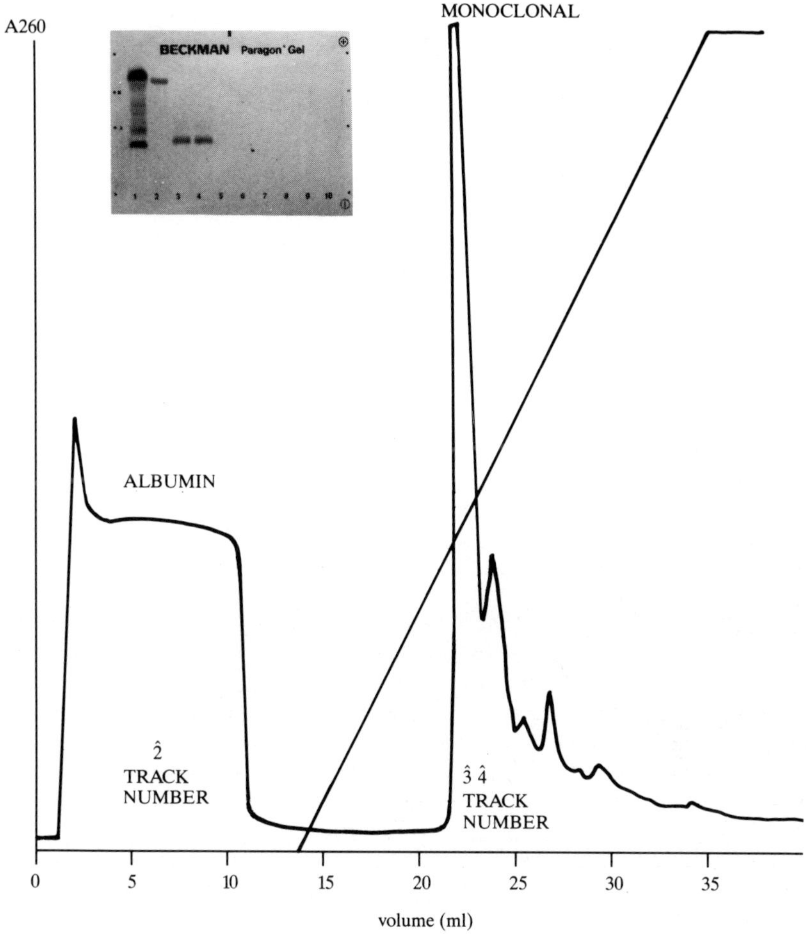

Figure 9 Cation exchange chromatography of ascitic fluid on Mono S. Conditions: 10ml of diluted ascites; column Mono S HR 5/5; buffer A 0.05M sodium acetate pH5.5; buffer B 0.05M sodium acetate pH5.5 with 0.5M sodium chloride; gradient 0% B for 14ml, 0-100% B in 21ml; flow rate 1ml/min

Inset Agarose gel electrophoresis of FPLC fractions. Track 1, unfractionated ascites; track 2, albumin containing fraction; tracks 3 and 4, monoclonal antibody containing fractions

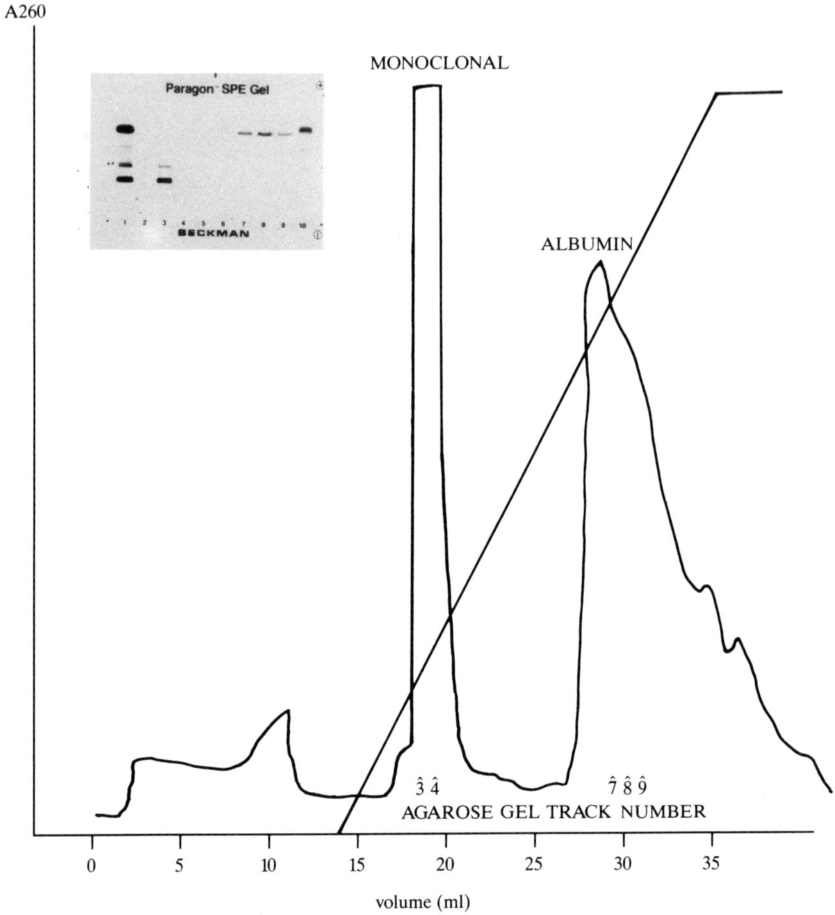

Figure 10 Anion exchange chromatography of ascitic fluid on Mono Q. Conditions: 10ml diluted ascitic fluid; column Mono Q HR 5/5; buffer A 0.005M Tris-HC1, pH 8.0; buffer B 0.2M Tris-HC1 pH 8.0 containing 0.2M sodium chloride; gradient 0% B for 14ml, 0-100% B in 21ml; flow rate 1ml/min.

Inset Agarose gel electrophoresis of FPLC fractions. Track 1, unfractionated ascites; track 3 and 4, monoclonal antibody containing fractions; tracks 7, 8 and 9, albumin containing fractions; track 10, normal mouse serum

freeze supernatants, ascites fluid or purified antibody in small aliquots. Such aliquots are best stored at −50°C or below. This inhibits enzymic proteolytic activity and helps to stabilize IgM isotypes. Alternatively, the antibody can be diluted 1:1 in glycerol and stored at −20°C; at this temperature the mixture remains liquid. Despite these precautions some monoclonal antibodies will still not be recoverable after freezing. Such antibodies should be diluted in 0.1M Tris-HCl buffer, pH 7.2, and stored at 4°C; Tris has considerable antibacterial activity.

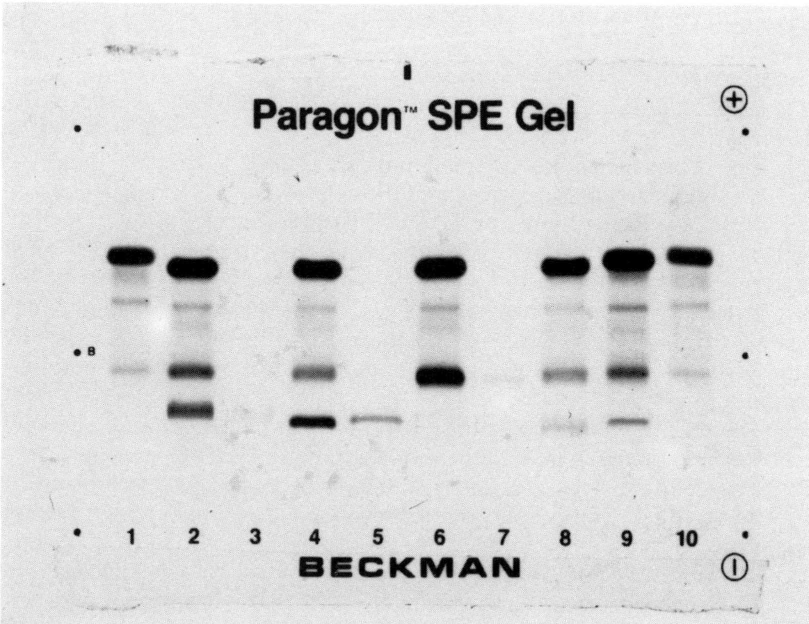

Figure 11 Agarose gel chromatography to monitor antibody purification. Track 1, normal mouse serum; track 2, ascitic fluid (IgM isotype); track 3, ascitic fluid (as track 2) purified on an ACA 22 column; track 4, ascitic fluid (IgG_{2a} isotype); track 5, ascitic fluid (as track 4) purified on a Protein-A sepharose column; track 6, ascitic fluid (IgG_1 isotype); track 7, ascitic fluid (as track 6) purified on a DEAE Tris-acryl column; track 8, ascitic fluid (IgG_{2b} isotype); track 9 ascitic fluid (IgG_3 isotype); track 10, normal mouse serum. Note that the rate of migration of the protein band cannot be used to determine the isotype.

Cryopreservation of hybridoma cells

The storage of cells in liquid nitrogen is an essential part of hybridoma technology. In the period immediately after fusion, the hybridoma cells are very fragile and readily undergo chromosomal rearrangement—even random chromosomal loss. It is essential to freeze the hybridoma cells as soon as possible after fusion. Sufficient cells may be available for freezing from the positive wells of the master plate. Thereafter, cells are frozen from positive wells following the first and each subsequent cloning step.

Materials

Cryopreservation medium: FCS containing 10% DMSO, stored at −20°C in 10ml aliquots (DMSO is toxic and does not require sterilizing before use)

Sterile plastics cryopreservation tubes (Nunc Plastics Ltd)

Cryopreservation vials which may be labelled on the cap are recommended, as they are more readily identified in the liquid nitrogen storage systems

Methods

1 The cell suspension is counted and centrifuged at 150g for 10min.

2 The supernatant is removed and the cells are resuspended in ice-cold cryopreservation medium to give a cell suspension of 5×10^5 - 1×10^7 cells/ml. 1ml of this cell suspension is then aliquotted into each cryopreservation tube.

3 Alternatively, cells from single wells of the master or cloning plates can be frozen. In this case, all the the medium in the well is carefully removed and the cells are resuspended in 1ml of cryopreservation medium.

4 The cells are kept on ice until frozen; at this temperature the DMSO is not as cytotoxic as it is at room temperature.

5 The cells must be frozen by a steady decrease in temperature of 1-2°C/min. Such conditions are achievable with commercial freezing apparatus.

6 Alternatively, the tubes are wrapped in paper tissues and enclosed in a polystyrene box with approximately ¾in-thick walls and with an approximate volume of 750cm^3. The box is sealed tightly with tape and placed in a −70°C freezer overnight. (The polystyrene boxes in which bottles of serum are often transported by commercial companies are ideal for this purpose.)

7 The frozen cells are quickly transferred to liquid nitrogen and stored either in the vapour phase or under the liquid.

Thawing frozen cells

1 The cryopreservation tube is removed from the liquid nitrogen and

quickly thawed by placing it in a universal containing sterile distilled water at 37°C.

2 The external surface of the tube is swabbed with 70% ethanol.

3 The cells are removed and quickly diluted to 20ml with RPMI containing 10% FCS.

4 The cells are centrifuged at 150g for 10min and the supernatant is discarded.

5 The cell pellet is resuspended in RPMI 1640 containing 20% FCS, to give a cell density of 1×10^5 - 5×10^5 cells/ml. As a rule of thumb, when thawing cells which have not been counted (ie from 96-well plates), the volume of medium used for resuspending should be twice that from which the cells were originally taken.

6 As hybridoma cells are very fragile after freezing, especially those cells stored immediately after fusion, it is recommended that such cells be cultured, at least initially, in the presence of feeder cells.

Depositing cells in ECACC

Culture collections provide invaluable support for scientists in all research programmes. For workers embarking on monoclonal antibody projects, the European Collection of Animal Cell Cultures (ECACC) can provide tangible assistance, from the supply of quality-controlled hybridoma fusion partners to eventual safe deposit (or distribution) of any hybridomas successfully produced.

The parent plasmacytoma cell lines, in many cases supplied to ECACC by the originating laboratory, are mycoplasma-tested before distribution. A collection of monoclonal-antibody-producing cell lines has been established and may already contain hybridomas which could be of assistance to workers wishing to develop immunodiagnostic assays.

After all the resources consumed in the production of important hybridomas, providing an alternative back-up source, as insurance against your own storage facilities failing, is a common-sense approach to long-term research programmes.

Additionally, once hybridomas are prepared, the burden of distribution of cell lines can be undertaken by ECACC, effecting a great saving in time and resources.

Details of all the faciltities provided by ECACC are given in a catalogue available, free of charge, from the following address.

The Curator
ECACC,
PHLS CAMR,
Porton Down,
Salisbury, SP4 OJG.
Telephone 0980 610391 ext 511

Chapter 5

Characterization

The extent to which the monoclonal antibodies produced need to be characterized will depend entirely on their intended use. For research purposes it may be necessary to make panels of antibodies directed against single antigens for epitope mapping, but for diagnostic reagents it is usually sufficient to demonstrate recognition of, and specificity to, the desired antigen under defined conditions.

In this chapter we propose a series of steps to characterize the antibody sufficiently to render it possible to assess its usefulness for diagnostic and commercial purposes.

Isotype analysis

The identification of the isotype of the monoclonal antibody is an essential prerequisite for purification and characterization. IgG isotypes are better for most assay systems, but the multivalency of IgM isotypes is sometimes preferred for assay systems based on agglutination.

IgG subclasses bind to Protein A with variable avidity (see page 48). Therefore, when purification is by Protein A affinity chromatography columns, or when Protein A is involved in the assay system, the subclass will need to be considered. The introduction of Protein G, with its broader specificity, may overcome some of these problems.

Several techniques are available to determine the antibody isotype, including immunodiffusion and ELISA.

Double immunodiffusion

Materials

0.5% (w/v) agar in distilled water
1% agar in PBS with 0.02% (w/v) sodium azide, pH 7.4
Stain—0.25% (w/v) Coomassie Blue R in destaining solution
Destaining solution—50% (v/v) methanol, 5% (v/v) acetic acid in distilled water

Method

1 Acid-cleaned glass microscope slides are placed on a level surface and precoated with 2.0ml of 0.5% agar at 56°C.

2 The agar is allowed to set. The slides are then left to dry overnight at room temperature. These precoated slides provide an adhesive support for the upper agar layer.

3 Precoated slides are coated with 3.5ml of molten 1% agar at 56°C.

4 After the agar has solidified, a series of 2-3mm-diameter wells are cut in the agar with a gel punch. In general, the wells are cut in a radial pattern, with one central well surrounded by a circle of six outer wells.

5 The agar plugs are carefully removed by vacuum aspiration or with a fine upturned needle.

6 10-100-fold concentrated culture supernatants, not ascites, are placed in the outer wells and undiluted polyclonal rabbit anti-mouse class-specific antisera are placed in the centre wells.

7 The slides are placed in a humidified box and incubated for 1-3 days at room temperature. Precipitin lines between an outer well and the centre well indicate the isotype of the immunoglobulin present.

8 Precipitin lines will be easier to view if the slides are stained. Excess protein is first removed by washing in several changes of PBS for 24h.

9 The agar is then covered with filter paper and left to dry overnight.

10 The filter paper is removed and the slides stained with staining solution for 1h at room temperature.

11 The stained slides are carefully destained and air-dried.

Note The double diffusion assay uses considerable quantities of expensive polyclonal anti-subclass antisera and is useless for ascites, where other mouse immunoglobulins will be present.

ELISA

The assay used in the monoclonal antibody screening test may be adapted to determine the isotype of the antibody bound to any solid phase antigen. For ELISA the method is as follows.

Method

1 Antigen is bound to the micro-ELISA plate and incubated with the supernatant as described on page 37.

2 After washing, the wells are incubated with 100μl of rabbit anti-mouse subclass-specific antisera (approximately 1:1,000 dilution in Tris-BSA buffer), for 2h at 37°C.

3 The wells are washed and incubated with 100μl of goat anti-rabbit IgG coupled to peroxidase (approximately 1:1,000 dilution in Tris-BSA buffer), for 30min at 37°C.

4 The plates are washed and incubated with substrate as previously described.

Alternative ELISA

Recently anti-subclass antisera coupled with biotin have become commercially available. Because of the amplification with streptavidin-biotinylated peroxidase, these antisera allow the determination of the isotype of monoclonal antibody in ascitic fluid, and even culture supernatant, by ELISA, without the need for the antibody to be first captured by antigen.

Method

1 Ascitic fluid is diluted (1:1,000) in coupling buffer (see page 37).

2 100μl of diluted ascitic fluid is added per well of an ELISA plate. Two wells are required for each subclass antisera to be tested. The plate is incubated overnight at room temperature.

3 The plate is washed three times with ELISA wash.

4 100μl of biotinylated goat anti-mouse subclass antisera (diluted 1:2,000 in ELISA diluent—anti-mouse IgG_1, IgG_{2a}, IgG_{2b}, IgG_3, IgM, IgA : Amersham Ltd) is added per well. The plate is incubated for 1h at 37°C.

5 The plates are washed three times with ELISA wash.

6 100μl of streptavidin-biotinylated-peroxidase complex (diluted 1:1,000) is added to each well and the plate is incubated for 15min at room temperature.

7 After washing, the bound peroxidase is detected with TMB substrate as previously described.

Antigenic specificity

We suggest that as many test systems as possible be used to characterize the conditions under which the monoclonal antibody will bind to the antigen. This type of information will be required before immuno-diagnostic assays can be established. It is not unlikely that antibodies which react in one test system, such as ELISA, will not react in another, such as immunofluorescence, or that those that react with purified antigens will not react with intact organisms. These anomalies can be ascribed to modification of the antigen, either during its preparation or during the assay, and confirm the importance of deciding beforehand which type of test system will be used and selecting the antibody accordingly.

Strain, species and genus cross-reactivity

Antigenic specificity will need to be defined at several levels. For immunodiagnostic purposes, the extreme specificity of monoclonal antibodies can be a distinct disadvantage. It is possible that strain-specific antibodies are produced with some antigens, and such antibodies are unlikely to be useful. In the first instance, the cross-reactivity of the antibody within strains of the same species should be established.

Cross-reactive antibodies can then be assessed against other members of the same genus. To this end, accessiblity to large collections of strains is essential. Additionally, it is important to check the specificity against the type strains, from recognized collections such as The National Collection of Type Cultures (NCTC, PHLS, 61 Colindale Avenue, London NW9 5TH), to which other workers will have access. Finally, the cross-reactivity with unrelated organisms should also be established.

It is possible that antibodies with partial cross-reactivity against strains of the same species will have serotype or serogroup specificity; in these cases, access to a range of previously serotyped strains is also important.

It is always wise to check that the antibody does not cross-react with host-tissue antigens, especially if it is intended for use in immunohistological systems. This type of cross-reaction is not necessarily non-specific, but may be due to common antigens such as carbohydrate groups.

Checking large numbers of strains may be a tedious procedure, and the development of a rapid test such as the immunodot test (see page 40) may be useful.

Antigen characterization

In most cases a very complex mixture of protein antigens will have been used to immunize the mouse. The specificity of the antibodies derived from that immunization will usually need to be defined down to the molecular level if these antibodies are to be diagnostically valuable. Even if a relatively pure protein was used to immunize the mouse it is necessary to confirm the nature of the antigen against which the antibody is directed.

Immunolabelling for electron microscopy

Visualizing the site at which the antibody binds to the organism can provide an indication of the nature of the antigen involved—for example, flagella or pili. Sometimes immunofluorescence studies may provide this information, but the electron microscope has much higher resolution and it may even be possible to distinguish between antibodies by differential labelling patterns.

With certain antigens—for example, pili, with their regular filamentous structure—antibody-binding sites can be visualized without any further enhancement, particularly when the antibodies are of the IgM isotype.

Method

1 Intact organisms or isolated pili or flagella are adsorbed onto copper grids coated with carbon-formvar.

2 Excess antigen is removed with the edge of a piece of torn filter paper, and the grid is inverted onto a droplet of monoclonal antibody (undiluted supernatant or ascites diluted 1:100 in PBS). The antibody droplet is formed by placing $10\mu l$ of fluid onto the hydrophobic surface of a small

sheet of GelBond film (Miles Research Labs), placed in a humidified chamber.

3 The grid is incubated overnight at 4°C, then washed by sequential inversion onto ten droplets of PBS followed by two drops of distilled water.

4 The grid is then negatively stained with 0.5% (w/v) phosphotungstic acid, pH 4.0.

Immunogold labelling In general, the site of binding of the antibody to the organism will need to be visualized by some electron-dense probe which can be more readily observed. Some investigators label their monoclonal antibody directly with colloidal gold particles, but this is usually unnecessary especially now that second antibodies labelled with various sizes of gold are commercially available.

Most, but not all, antigens will retain their antigenicity after a light fixation. Such fixation will help to maintain the morphological integrity of the organism during the rest of the procedure, but it must be remembered that reactions dependent on three-dimensional configuration of the antigen may not survive such treatment. In our experience, non-specific binding is reduced by incubating the primary antibody with the organism in solution. After washing, the cells are then placed on a grid for incubation with the gold probe.

Method

1 If the cells are to be fixed, they are washed free of proteinaceous culture material and treated with 0.1% (v/v) glutaraldehyde in 0.1M phosphate buffer, pH 7.2, for 10min at room temperature, before washing and storing in 0.1M phosphate buffer at 4°C.

2 Fixed or unfixed cells are suspended in either ascitic fluid (1:100 dilution in PBS) or culture supernatant and incubated on a rotation stirrer for 1h at room temperature.

3 The organisms are washed three times in PBS by centrifugation in a microcentrifuge for 2min and resuspended in a small volume of PBS (volume dependent on the number of bacteria), and the bacteria are absorbed onto a copper electron microscope grid coated with carbon-formvar for a few minutes. The excess fluid is removed with the edge of a piece of torn filter paper.

4 The grid is then immediately inverted onto a 20μl droplet of rabbit anti-mouse IgG coupled to 5-15nm gold particles (Janssen Phamaceuticals Ltd) placed on the hydrophobic side of a sheet of GelBond film. 15nm gold is easily seen at low magnification, but can cause stearic hindrance to labelling. 5nm gold will distinguish individual antigenic molecules.

5 The grid is incubated with the probe at 4°C overnight in a humidified chamber, and then washed by inversion on to ten droplets of PBS and two droplets of distilled water.

6 After washing, the grid is stained with 0.5% (v/v) phosphotungstic acid.
7 It is recommended that polyclonal antisera at 1:100 dilution be used as a positive control and an unrelated monoclonal antibody as a negative control.

Note Some monoclonal antibodies will not be visualized by immunolabelling, even with unfixed bacteria. The reason for this is usually that the epitopes, against which the antibodies are directed, are 'hidden', precluding binding to the antibody.

Defining the chemical nature of the antigen

Some indication of the chemical nature of the antigen can be obtained by the loss of antigenicity after various treatments. Protein antigens will be sensitive to treatment with a protease such as trypsin or pronase. Sensitivity to heat treatment (60°C or 100°C) also indicates a protein antigen. Loss of antigenicity after treatment with periodate suggests the antigen is a carbohydrate (although tyrosine and tryptophan residues are also sensitive) or a glycoprotein.

Defining the antigen at the molecular level

There are two complementary techniques for characterizing protein antigens—Western blotting and radioimmunoprecipitation.

In Western blotting the antigen is first separated into its components by electrophoretic techniques, such as sodium dodecyl sulphate—poly acrylamide electrophoresis (SDS-PAGE). These bands are then transferred onto nitrocellulose paper. The nitrocellulose can then be incubated with the antibody and bound antibody can be detected by immunostaining techniques.

In the radioimmunoprecipitation technique the antigens are first radiolabelled and then incubated with the antibody. Antibody is then immunoprecipitated by Sepharose beads coupled to protein-A or anti-mouse IgG. Material bound to the beads can then be separated by SDS-PAGE and radiolabelled polypeptide bands be detected by autoradiography.

These two techniques do not necessarily give identical results, but should be used to complement each other.

SDS-polyacrylamide gel electrophoresis

Antigenic mixtures, such as proteins, need to be separated by SDS-PAGE for analysis. Which electrophoretic system to use will depend on the antigenic mixture required to be separated, and should be established before monoclonal antibody characterization is embarked on. Because the electrophoretic conditions need to be established for each antigen, detailed description of SDS-PAGE will not be included. However, in our experience a 10-25% gradient gel system using the discontinuous buffer system of Laemmli (1970) appears to separate the polypeptide components of most bacterial and viral antigens. This system will adequately separate polypeptides from 200-10kDa with reasonable resolution. Many excellent technical books are available to provide information on SDS-PAGE techniques (see *Recommended Reading*, page 82). Non-reducing systems, which may eliminate antigenic loss, and two-dimensional electrophoresis should also be considered.

Western blotting

Proteins separated by SDS-PAGE can be transferred electrophoretically onto a paper which is then incubated with antibodies in order to determine which proteins are antigenic. Until recently, this electrophoretic transfer had been done by sandwiching the gel with the nitrocellulose between plastics sheets in a large tank, as described by Towbin (1979). The use of recently developed semi-dry blotting systems (Kyhse-Andersen 1984) is considerably more efficient, using less expensive reagents, saving time and generating far less antigen-damaging heat. Semi-dry blotting apparatus is commercially available (Ancos Ltd and Sartorius Ltd) and comprises two horizontal graphite sheets encased in a plastics base and lid, between which the gel and the nitrocellulose are sandwiched.

It must be remembered that not all antigens will retain their antigenicity under the conditions of SDS-PAGE separation.

Antibodies directed against primary sequences are most likely to give positive results with this technique. Sometimes the polypeptides can be renatured by incubation of the gel in renaturing solution. Otherwise, PAGE can be carried out in non-denaturing conditions.

Materials

Whatman filter paper No 3 (cut to the size of the gel)
Whatman filter paper No 1 (as above)
Nitrocellulose paper (as above). Nylon paper is not recommended for this system
Cathodic buffer: 0.30% (w/v) Tris, 0.523% (w/v) 6-amino-n-hexanoic acid, 20% (v/v) methanol
Anodic buffer No 1: 3.63% (w/v) Tris, 20% (v/v) methanol
Anodic buffer No 2: 0.303% (w/v) Tris, 20% (v/v) methanol
(The cathodic and anodic buffers should be made in small quantities and stored for no more than 2 weeks at 4°C)
TBS: 0.242% (w/v) Tris, 2.924% (w/v) sodium chloride, buffered to pH 7.5 with HC1
TTBS: as for TBS, with 0.05% (v/v) Tween 20
Blocking buffer: 3% (w/v) gelatin dissolved in TBS at 37°C
Incubation buffer: 1% (w/v) gelatin dissolved in TTBS at 37°C
Peroxidase substrate: 60mg 4-chloro-1 naphthol dissolved in 20ml ice-cold

methanol and diluted in 100ml TBS containing 60μl hydrogen peroxide. Make up fresh each time (4-chloro-1-naphthol should be stored at −20°C)
Protein-staining solution: 0.05% (w/v) Coomassie Blue R, 20% isopropanol, 10% glacial acetic acid
Destaining solution: 5% (v/v) isopropanol, 5% (v/v) glacial acetic acid

Method

1 All gels, blotting paper and nitrocellulose are handled with plastics gloves.
2 The plates are prised open, exposing the run gel but causing as little distortion to the stacking gel as possible.
3 The separating gel is cut into appropriate slices with a razor blade, using the stacking gel as a guide, care being taken to avoid fragmenting the gel.
4 The stacking gel is carefully removed.
5 The lower, anode graphite plate of the semi-dry blotter is rinsed with distilled water. (In some systems the anodic plate is the upper plate, so the method is altered accordingly.)
6 Three sheets of No 3 filter paper are soaked in anodic buffer No 1 and layered onto the anode plate. It is important that no air bubbles be trapped between the layers of gel, nitrocellulose or filter paper.
7 Three sheets of No 1 filter paper are soaked in anodic buffer No 2 and layered onto the previous sheets.
8 A sheet of nitrocellulose, soaked in distilled water, is layered onto the filter papers.
9 The gel is arranged on the nitrocellulose. If the gloves are kept wet, this reduces the risk of tearing the gel.
10 Three sheets of No 1 filter paper are soaked in cathodic buffer and layered onto the gel.
11 The last four stages (7, 8, 9 and 10) comprise a stacking unit (see Figure 12, page 68). Up to six stacking units can be arranged in the semi-dry blotter. Each stacking unit must be separated by a sheet of dialysis membrane (Dako Ltd) soaked in distilled water. This dialysis membrane is not required if only one stacking unit is used.
12 Three sheets of No 3 filter paper are soaked in cathodic buffer and layered onto the final stacking unit.
13 The upper graphite plate is rinsed in distilled water and placed on top of the filter papers.
14 The power supply is connected and run at 0.8mA per cm^2 of gel at constant current for 1-2h.
15 The power supply is disconnected, and the lid and filter papers are carefully removed, to expose the gel slices.
16 The nitrocellulose is carefully cut around the gel with a scalpel blade.
17 The gel is removed and stained for 3h at 37°C with Coomassie Blue R, to confirm transfer of the proteins, and then destained with destaining

solution.

18 Meanwhile, the upper surface of the nitrocellulose is marked with indelible ink, the strips being numbered as necessary.

19 The nitrocellulose blots are incubated in blocking buffer at room temperature for 1h.

20 These blocked nitrocellulose blots may be wrapped in cling film and stored at −20°C for several weeks.

Immunostaining of blots It is important that the blots do not dry at any stage during the immunolabelling: this will give high background labelling. If the blots have been frozen, ensure that the gelatin has melted before staining by preincubation at 37°C for 30min.

Method

1 Blocked nitrocellulose blots are incubated with primary antibody for 1h at room temperature on an orbital stirrer. Monoclonal antibody supernatant can be used undiluted with 20mM HEPES to maintain a

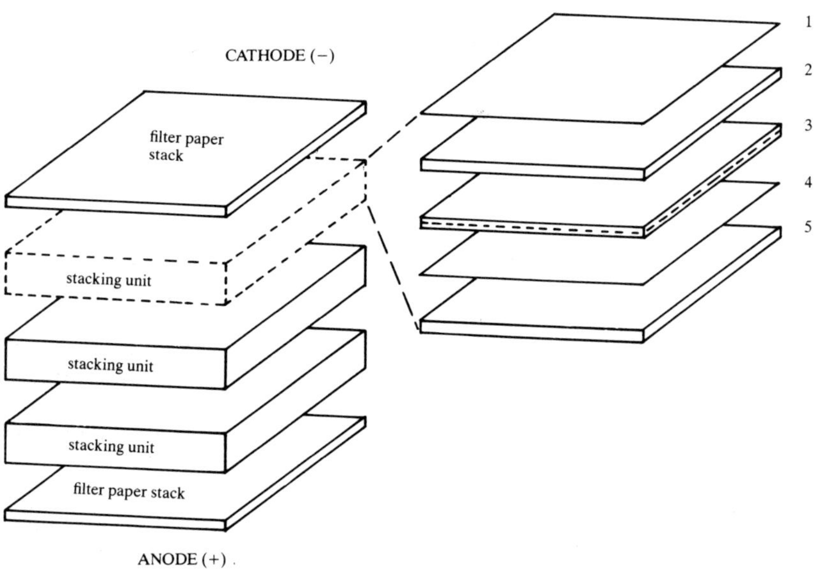

Figure 12 Stacking unit. 1 dialysis membrane, 2 filter paper stack, 3 SDS-polyacrylamide gel, 4 nitrocellulose paper, 5 filter paper stack

suitable pH. Ascites and hyperimmune antisera should be diluted 1:100 with incubation buffer. Enough antibody should be used to adequately cover the nitrocellulose.

2 The blots are washed for 30min with five changes of TTBS.

3 The blots are then incubated with secondary antibody for 1h at room temperature, with shaking. Antibodies conjugated to peroxidase (or alkaline phosphatase) and sold especially for blotting are recommended, as they have a high titre and are screened for non-specific binding. Conjugated antibodies should be diluted with an equal volume of glycerol and stored at −20°C to prevent freeze-thaw problems.

4 The blots are washed for 30min with six changes of TTBS, and then washed for 5min with two changes of TBS.

5 The blots are incubated with substrate until the colour is sufficiently developed, washed briefly in distilled water, dried immediately and stored in the dark.

Notes

1 If several antibodies need to be investigated against the same antigen, a blank sample comb can be used for the SDS-PAGE gel. The gel can then be blotted whole and the nitrocellulose cut into 20-30 strips for immunostaining.

2 In order to reduce the amount of antibody required for the incubation steps, troughs, sufficient to take single nitrocellulose strips (1ml of antibody is sufficient in a trough of 1cm × 14cm), can be cut in Perspex sheets. Such troughs are also available from BioRad Ltd. Alternatively, apparatus which exposes strips of uncut nitrocellulose for incubation is commercially available.

3 Amino ethyl carbazole may be used as an alternative substrate for peroxidase, instead of 4-chloro naphthol. The substrate solution is made as described previously for immunohistological staining (see page 35) and produces a reddish-brown precipitate on the blot. The reaction is stopped, once sufficient colour has developed, by washing in distilled water.

4 ^{125}I-labelled protein-A or ^{125}I-labelled anti-mouse IgG (1×10^6cpm/ml) can be used instead of enzyme conjugated second antibodies, but have little advantage and require specialized equipment and facilities.

5 Because the gel expands after staining, it is difficult to determine the molecular weights of immunolabelled bands. Some prestained markers are commercially available (BioRad Ltd; Amersham International), but do not allow accurate determination of molecular weight, although they are excellent for marking tracks on the blot and for checking that the gel has run as expected. Biotinylated molecular weight standards are now available (BioRad Ltd) which can be visualized by immunostaining with avidin conjugated to peroxidase. It is also possible to stain transferred proteins on nitrocellulose with 0.5% (w/v) Amino Black in 50% (v/v)

methanol and 10% (v/v) acetic acid for 5min. The stained paper is destained with 50% (v/v) methanol and 10% (v/v) acetic acid until the required background is obtained. However, this method of staining is very insensitive.

6 The inability of monoclonal antibodies to bind to antigens in Western blots may be due to either the poor transfer of the antigen onto the nitrocellulose or the loss of antigenicity due to the SDS-PAGE separation, which destroys three-dimensional configuration.

7 The inefficient transfer of some bacterial outer membrane proteins can be improved by the inclusion of 0.1% (w/v) SDS in the transfer buffers.

8 Preincubation of the gel in a renaturing solution of 10mM Tris-HC1 buffer, pH 7.0, containing 50mM sodium chloride, 4M urea and 0.1mM dithiothreitol, for 1h at room temperature before blotting may renature some polypeptides.

Radioimmunoprecipitation

Whole bacteria or cells may be radioiodinated by use of techniques which label only the surface exposed proteins and, it is hoped, do not allow the isotope to penetrate the cell and so preclude labelling of the intracellular components. Lactoperoxidase is generally used to catalyse the iodination of cell surface proteins, as this procedure is very mild.

Chloramine T and Iodogen (see opposite) are both excellent means of radioiodinating isolated proteins, giving significantly higher specific activities than the lactoperoxidase technique; but because they are oxidizing agents, the reaction is potentially damaging to the proteins.

^{35}S may also be used to label protein antigens, although the difference in detection systems makes this method less sensitive.

All of these techniques require access to radioactive facilities. This book cannot undertake the detailed description of the techniques involved in radioiodination of the antigens, but such details are available elsewhere (see the Recommended reading list).

Once the antigens—for example, the surface proteins of a bacterium—are radiolabelled, the sample must be solubilized in a suitable detergent. Any insoluble material must be removed before the antigen is mixed with the monoclonal antibody. The antibody-antigen complexes formed are selectively removed by Protein-A coupled to Sepharose beads. The beads are washed free of any unbound radiolabelled antigen. Bound antibody-antigen complexes are solubilized in dissociation buffer and run on an SDS-PAGE gel. The radiolabelled bands are detected by autoradiography.

Materials for ^{125}I-surface labelling by the lactoperoxidase technique

PBS: 0.8% (w/v) sodium chloride, 0.02% (w/v) potassium chloride, 0.115% (w/v) disodium hydrogen phosphate, 0.02% (w/v) potassium

dihydrogen phosphate, 0.01% (w/v) calcium chloride, 0.01% (w/v) magnesium chloride
Lactoperoxidase: 1mg/ml (Sigma Ltd)
(^{125}I)-Na (Amersham International)
0.02M hydrogen peroxide
5mM cysteine

Method

1 Bacteria are washed twice with PBS by centrifugation.
2 Approximately 1×10^9 bacteria are suspended in 500μl of PBS.
3 5μl of lactoperoxidase solution and 50μCi of (^{125}I)-Na are added to the bacteria.
4 Three aliquots of 50μl of hydrogen peroxide solution are added at 2.5min intervals.
5 The reaction is stopped after 10min by the addition of 10ml 5mM cysteine.
6 The bacteria are washed three times by centrifugation and the specific activity is determined.

Materials for ^{125}I surface labelling by the Iodogen technique

1,3,4,6-tetrachloro-3a,6a-diphenyl glycuril (Iodogen: Pierce Chemical Co); 10μg Iodogen dissolved in chloroform used to coat the bottom of a small glass tube. The chloroform is vaporized under a gentle stream of nitrogen. The tubes are coated just before use. (Or Iodobeads—Pierce Chemical Co—can be used.)
(^{125}I)-Na (Amersham International)
PBS: as above
10^{-5}M potassium iodide

Method

1 Approximately 2.5×10^8 bacteria are harvested and washed twice in PBS by centrifugation.
2 The bacteria are resuspended in 50μl of PBS and placed in the Iodogen-coated tube or in a siliconised tube containing one Iodobead.
3 5μl potassium iodide solution and 500μCi (^{125})-I are added to the tube and incubated for 10min at room temperature.
4 The bacteria are quickly removed from the tube and washed three times in PBS by centrifugation.
5 The specific activity is determined.

Materials for radioimmunoprecipitation technique

Radiolabelled antigens solubilized for 1h at 40°C in 0.1% (w/v) SDS, 1% (v/v) Empigen BB in PBS and centrifuged at 10,000g for 1.5h to remove any insoluble material
Protein-A-Sepharose 4B beads (Sigma Chemicals Ltd) (5mg) swollen at

4°C in 100μl of PBS containing 1mg of BSA; or anti-mouse IgG-Agarose (Sigma Chemicals Ltd)

Method

1 80μl of ^{125}I-labelled material (the dilution will depend on the specific activity of the relevant antigens, but 1-5 × 10^6cpm of labelled material should be sufficient) is mixed with 40μl of monoclonal antibody at 4°C on a rotating mixer for 20min.

2 40μl of swollen Protein-A-Sepharose (or 25μl of anti-mouse IgG-Agarose) is added and the mixture is rotated for a further 60min at room temperature.

3 The immune complexes bound to the beads are sedimented in a microfuge at 12,000g for 30s and washed five times in PBS at 4°C.

4 The washed beads are resuspended in 50μl SDS-PAGE dissociation buffer and boiled at 100°C for 5min. The supernatant is loaded onto a SDS-PAGE gel and separated by the standard technique adopted.

5 After electrophoresis, the gel is stained, destained and dried.

6 The dried gel is placed next to a sheet of X-ray film (X-Omat AR film: Kodak Ltd) in a X-ray cassette containing an enhancing screen at −70°C until sufficient of the radiolabelled bands is clearly seen.

Chapter 6

Immunodiagnosis

It is assumed throughout this book that the reader wishes to produce monoclonal antibodies as in house immunodiagnostic reagents. Thus, after some considered thought, several working weekends and a few muttered prayers to the appropriate Gods, a bottle of promising monoclonal antibody occupies pride of place in the refrigerator. This antibody needs then to be evaluated for antigen detection purposes.

Antigen detection techniques can be broadly divided into two types—antigen immunolabelling where the antigen is visualized by labelling the antibody—eg immunofluorescence; and antigen capture, where minute concentrations of antigen are trapped and this captured antigen is detected—eg latex agglutination.

Indirect antigen immunolabelling

In this technique the monoclonal antibody is incubated with the antigen—for example, bacteria on slides. The bound monoclonal antibody is then visualized after incubation with an anti-mouse IgG coupled to a label such as peroxidase or FITC. Such enzyme and fluorochrome-labelled anti-mouse IgG probes are readily and cheaply available. Indirect labelling of the antibody has the advantage that it does not require manipulation of the monoclonal antibody and that the additional step allows some amplification of the signal, but the disadvantage of requiring yet another incubation step.

Direct labelling

In this case the monoclonal antibody is coupled directly to an appropriate label (enzyme, radioisotope or fluorochrome). The major advantage of this procedure is that only a single incubation step is involved in the detection of the antigen. Its disadvantage is that all coupling procedures affect the activity of the antibody. The larger the label and the higher the number of molecules coupled to the antibody, the less antibody activity is retained and there is a probability of stearic hindrance of the antigen binding. It is, therefore, essential to check the antibody activity after coupling to any label. Some of the smaller labels, such as ^{125}I and biotin, may overcome this problem. The stability of the antibody during the coupling procedure will depend on the physical properties of that antibody: methods which work well for one antibody may completely denature another.

The following labels may be used.

Enzymic	horseradish peroxidase alkaline phosphatase ß-glucuronidase
Radioisotopic	^{125}I ^{35}S
Fluorescent	FITC Texas Red Rhodamine
Other	Biotin

Some commercial companies will provide an antibody labelling service, which could be cost-effective if biochemical expertise in the laboratory is limited.

Antigen capture

The antigen is captured, from a complex mixture, by an antibody bound to a solid support. Once captured, the antigen is detectable either visually or immmunologically. Obviously, for effective capturing of the antigen, the antibody should be exposed over as large a surface area as possible.

The simplest antigen-capture technique involves the absorption of the monoclonal antibody onto suitable latex beads. On mixing with the clinical material, the antigen is bound and cross-links the beads, causing an agglutination which is visible.

This type of system has been used in some rotavirus kits. Unfortunately, few monoclonal antibodies allow effective latex agglutination by antigens. Alternatively, labelled probes are used to detect captured antigen. In these systems the monoclonal antibody can be used either as the capture antibody or as the detecting antibody.

As the binding of monoclonal antibodies to solid supports such as ELISA plates can significantly reduce their activity, it is usual to use polyclonal antibodies as the capture system and a labelled monoclonal antibody as the more specific detection system.

It is possible to avoid directly labelling the monoclonal antibody by using an indirect system, an anti-mouse IgG coupled to the label, but this is cumbersome and requires the polyclonal to be derived from a different species than the monoclonal antibody.

Commercialization

Anyone who has purchased monoclonal antibodies will be impressed by the potential value of even the most obscure antibody. However, before your supply of antibody can be converted into a much-needed new ultracentrifuge, a saleable product must be offered. The development of immunodiagnostic kits is a specialized area of research, and requires marketing skills which we do not recommend any research or routine laboratory to attempt. A number of commercial companies have recently developed interests in the marketing of immunodiagnostic kits and reagents for infectious diseases. A scan of the advertising literature and Linscott Directory will provide some indication of the companies worth contacting. Before approaching any commercial company, it is wise to establish:

- the ownership of the hybridoma (this is especially of concern when research grants are involved)
- whether the clone or just the antibody can be offered for sale
- the financial terms to be considered—ie one off payment and/or royalties (check current rate with other laboratories)
- whether a sole agent agreement is wanted or not.

Additionally, the companies will want the following information:

- all of the known characteristics of the antibody
- the stability of the hybridoma
- an assurance that adequate quantities of ascites will be available
- a sample of ascities for investigation
- an assurance that adequate stocks of the hybridoma are stored in secure liquid nitrogen tanks.

References

Bennet FC, Yeoman LC. An improved procedure for the "Dot immumobinding" analysis of hybridoma supernatants. *J Immunol Methods* 1983, **61**, 201-7.

Borrebaeck CAK, Möller SA. In vitro immunization. Effect of growth and differentiation factors on antigen-specific B cell activation and production of monoclonal antibodies to autologous antigens and weak immunogens. *J Immunol* 1986, **136**, 3710-15.

Boss BD. An improved *in vitro* immunization procedure for the production of monoclonal antibodies. *Methods Enzymol* 1986, **121**, 27-33.

Chapel HM, August PJ. Report of nine cases of accidental injury to man with Freund's complete adjuvant. *Clin Exp Immunol* 1976, **24**, 538-41.

Colwell DE, Michalek SM, McGhee JR. Method for generating a high frequency of hybridomas producing monoclonal IgA antibodies. *Methods Enzymol* 1986, **121**, 42-51.

Engvall E, Perlmann P. Enzyme-linked immunosorbent assay (ELISA). Quantitative assay of immunoglobulin G. *Immunochemistry* 1971, **8**, 871-4.

Filip C, Fletcher G, Wulff JL, Earhart CF. Solubilization of the cytoplasmic membrane of *Escherichia coli* by the ionic detergent sodium-lauryl sarcosinate. *J Bacteriol* 1973, **115**, 717-22.

Galfrè G, Milstein C. Preparation of monoclonal antibodies: strategies and procedures. *Methods Enzymol* 1981, **73**, 3-46.

Galfrè G, Milstein C, Wright B. Rat × rat hybrid myelomas and a monoclonal anti-Fd portion of mouse IgG. *Nature* 1979, **277**, 131-3.

Goka AKJ, Farthing MJG. The use of 3,3′,5,5-tetramethylbenzidine as a peroxidase substrate in microplate enzyme-linked immunosorbent assay. *J Immunoassay* 1987, **8**, 29-41.

Goldstein MA, Tauraso NM. Effect of formalin, *p*-propiolactone, merthiolate and ultraviolet light upon influenza virus infectivity, chicken cell agglutination, hemagglutination and antigenicity. *Appl Microbiol* 1970, **19**, 290-4.

Habermann E, Goretzki K. Monoclonal antibodies against tetanus toxin and tetanus toxoid. In: Macario AJL, Conway de Macario E, eds. Monoclonal antibodies against bacteria; Vol 1. Orlando: *Academic Press* 1985, 191-205.

Handman E, Remington JS. Serological and immunochemical characterization of monoclonal antibodies to *Toxoplasma gondii*. *Immunology* 1980, **40**, 579-88.

Heckels JE, Virji M. Monoclonal antibodies against gonococcal pili: uses in the analysis of gonococcal immunochemistry and virulence. In:

Macario AJL, Conway de Macario E, eds. Monoclonal antibodies against bacteria; Vol 1. Orlando: *Academic Press*, 1985, 1-35.

Himmelweilt F. Serological responses and clinical reactions to influenza virus vaccines. *Br Med J* 1960, **ii**, 1690-4.

Hirsch RL, Griffin DE, Winkelstein JA. Host modification of sindbis virus sialic acid content influences alternative complement pathway activation and virus clearance. *J Immunol* 1981, **127**, 1740-3.

Hurn BA and Chantler SM. Production of reagent antibodies. *Methods Enzymol* 1980, **70**, 104-42.

Jann K, Westphal O. Microbial polysaccharides. In: Sela M, ed. The Antigens 3. New York: *Academic Press* 1975, 1-125.

Johnson AM, McNamara PJ, Neoh SH, McDonald PJ, Zola H. Hybridomas secreting monoclonal antibody to *Toxoplasma gondii*. *Aust J Exp Biol Med Sci* 1981, **59**, 303-6.

Kilmartin JV, Wright B, Milstein C. Rat monoclonal antitubulin antibodies derived by using a new non-secreting rat cell line. *J Cell Biol* 1982, **93**, 576-82.

Köhler G, Milstein C. Continuous cultures of fused cells secreting antibody of predefined specificity. *Nature* 1975, **256**, 495-7.

Köhler G, Milstein C. Derivation of specific antibody-producing tissue culture and tumor lines by cell fusion. *Eur J Immunol* 1976, **6**, 511-19.

Komisar JL, Fuhrman JA, Cebra JJ. IgA-producing hybridomas are readily derived from gut-associated lymphoid tissue. *J Immunol* 1982, **128**, 2376-8.

Kyhse-Andersen J. Electroblotting of multiple gels: a simple apparatus without buffer tank for rapid transfer of proteins from polyacrylamide to nitrocellulose. *J Biochem Biophys Methods* 1984, **10**, 203-9.

Laemmli UK. Cleavage of structural proteins during the assembly of the head of the bacteriophage T4. *Nature* 1970, **227**, 680-5.

Lennette EH, Schmidt NJ, eds. Diagnostic procedures for viral, rickettsial and chlamydial infections. 5th ed. Washington: *American Public Health Association*, 1979.

Littlefield JW. Selection of hybrids from matings of fibroblasts in vitro and their presumed recombinants. *Science* 1964, **145**, 709-10.

Miura T, Mizushima S. Separation by density gradient centrifugation of two types of membranes from spheroplast membrane of *Escherichia coli* K12. *Biochim Biophys Acta*, 1968, **150**, 159-61.

Morgan DO, Moore DM, McKercher PD. Vaccination against foot-and-mouth disease. *Prog Clin Biol Res*, 1980, **47**, 169-78.

Oguma K, Syuto B, Kubo S and Iida H. In: Macario AJL, Conway de

Macario, eds. Monoclonal antibodies against bacteria; Vol 2. Orlando: *Academic Press* 1985, 160-.

Osborn MJ, Munson R. Separation of the inner (cytoplasmic) and outer membranes of gram-negative bacteria. *Methods Enzymol* 1974, **31**, 642-53.

Pertoft H. Purification of herpes simplex virus using Percoll. *Separation News* 1980, No. 3.

Potter M, Robertson CL. Development of plasma-cell neoplasms in BALB/C mice after intraperitoneal injection of paraffin-oil adjuvant, heat-killed staphylococcus mixtures. *JNCI* 1960, **25**, 847-61.

Reading CL. *In vitro* immunization for the production of antigen-specific lymphocyte hybridomas. *Methods Enzymol* 1986, **121**, 18-27.

Redfield DC, Richman DD, Oxman MN. Psoralen inactivation of influenza and herpes simplex viruses and virus infected cells. *Infect Immun* 1981, **32**, 1216-26.

Sato H. Monoclonal antibodies against *Clostridium perfringens* θ toxin (perfringolysin O). In: Macario AJL, Conway de Macario E, eds. Monoclonal antibodies against bacteria; Vol 3. Orlando: *Academic Press*, 1986, 203-28.

Schulman M, Wilde CD, Köhler G. A better cell line for making hybridomas secreting specific antibodies. *Nature* 1978, **276**, 269-70.

Shone C, Wilton-Smith P, Appleton N, Hambleton P, *et al.* Monoclonal antibody-based immunoassay for type A *Clostridium botulinum* toxin is comparable to the mouse bioassay. *Appl Environ Microbiol* 1985, **50**, 63-7.

Silverman M, Simon MI. Bacterial flagella. *Annu Rev Microbiol* 1977, **31**, 397-419.

Smith E, Roberts K, Butcher GW, Galfrè G. Monoclonal antibody screening 2: methods using antigens immobilized on nitrocellulose. *Anal Biochem* 1984, **138**, 119-24.

Spitz M. "Single-shot" intrasplenic immunization for the production of monoclonal antibodies. *Methods Enzymol* 1986, **121**, 33-41.

Svennerholm A-M, Wikström M, Lindholm L, Holmgren J. Monoclonal antibodies and immunodetection methods for *Vibrio cholerae* and *Escherichia coli* enterotoxins. In: Macario AJL, Conway de Macario E, eds. Monoclonal antibodies against bacteria; Vol 3. Orlando: *Academic Press* 1986, 77-97.

Thompson NE, Bergdoll MS, Meyer RF, Bennet RW, Miller L, Macmillan JD. In: Macario AJL, Conway de Macario E, eds. Monoclonal antibodies against bacteria; Vol 2. Orlando: *Academic Press* 1985.

Tint H, Stone J. An adenovirus vaccine potency test. *J Immunol* 1961, **86**, 253-6.

Towbin H, Staehelin T, Gordon J. Electrophoretic transfer of proteins from polyacrylamide gels to nitrocellulose sheets: Procedure and some applications. *Proc Natl Acad Sci USA* 1979, **76**, 4350-4.

Recommended reading

1 Johnstone AP, Thorpe R. Immunochemistry in practice. St. Louis: Mosby, 1982.

2 Hudson L, Hay FC. Practical immunology. 2nd ed. Oxford: Blackwell Scientific Publications, 1981.

3 Mishell BB, Shiigi SM, eds. Selected methods in cellular immunology. New York: WH Freeman, 1980.

4 Di Sabato G, Langone JJ, Van Vunakis H, eds. Immunochemical techniques. Part G. Separation and characterization of lymphoid cells. Orlando: Academic press, 1984. (Colowick SP, Kaplan NO, eds. Methods in enzymology; vol 108.)

5 Langone JJ, Van Vunakis H, eds. Immunochemical techniques. Part I. Hybridoma technology and monoclonal antibodies. Orlando: Academic Press, 1986. (Colowick SP, Kaplan NO, eds. Methods in enzymology; vol 121.)

6 Weir DM, Blackwell, CC, Herzenberg LA, Herzenberg LA, eds. Handbook of experimental immunology. Vol 4. Applications of immunological methods in biomedical sciences. Oxford: Blackwell Scientific Publications, 1986.

7 Voller A, Bartlett A, Bidwell D, eds. Immunoassays for the Eighties. Baltimore: Univ Park Press, 1981.

8 Macario AJL, Conway de Macario E, eds. Monoclonal antibodies against bacteria. Vol 1-3. Orlando: Academic Press, 1985-86.

9 Goding JW. Monoclonal antibodies: principles and practice. Orlando: Academic Press, 1983.

10 Wicher K ed. Microbial antigen diagnosis. Vol 1 & 2. Boca Raton: CRC Press Inc, 1987.

Acknowledgements

We are most grateful to Ann Williams, Diane Harwood and Susan Cima for their generous help with Figures 1, 2, 4 and 5. Figure 3 was kindly supplied by The European Collection of Animal Cell Cultures, Porton Down.

Index

PHLS publications

The Public Health Laboratory Service (PHLS) is a statutory organisation concerned with the diagnosis, surveillance and control of infectious diseases in England and Wales. Since the early 1970s the PHLS has produced books and serial publications for medical microbiologists and others concerned with the laboratory diagnosis, transmission and prevention of communicable diseases.

The PHLS publications listed on these pages are available to interested purchasers worldwide. A full Publications Catalogue is available from the address at the end of this page.

The PHLS as a publisher

The PHLS acts in the same way as a normal commercial publisher. The Press and Publications Officer is authorised to consider for publication any suitable manuscript on medical microbiology or associated fields. The PHLS pays its authors and editors. Authors of PHLS publications are often on the staff of the PHLS, but this is not a necessary requirement. Just like any other publishing house, the PHLS considers all manuscripts on their merits.

The only way in which the PHLS differs from a commercial publishing house is in its terms of supply.

Terms of supply

PHLS publications are supplied worldwide from the UK on a **strictly payment with order basis**. This is to minimise administrative costs and thus prices. All book prices quoted are inclusive of postage and packing costs. A receipt is supplied with all orders filled.

Wherever possible, orders should be accompanied by a cheque or money order made out in pounds sterling which can be drawn on a bank within the UK. Where overseas customers cannot do this, cheques should be made out in the local currency at the current rate of exchange against the sterling price.

All cheques and money orders should be made payable to the Public Health Laboratory Service.

Servicing orders

The PHLS aims to service all routine orders within 48 hours of receipt. Under normal circumstances a UK customer can therefore expect to receive publications within days of order. Overseas customers' orders are dispatched by second class airmail — serial publications as well as books.

Address for orders and correspondence

All orders, and correspondence regarding these orders, should be addressed to PHLS Supplies Department, 61 Colindale Avenue, London NW9 5DF, UK (telephone 01-200 1295). All other enquiries and editorial correspondence should be sent to the Press and Publications Officer at the same address.

PHLS Serial publications

Sample copies of both publications are available.

PHLS Microbiology Digest

ISSN 0265-3400, published quarterly, one volume per annum, approximately 80 A4 pages per volume. Free to *PHLS Library Bulletin* subscribers, also available separately at £10 per annum (1989).

PHLS Microbiology Digest contains mainly brief and up-to-date reviews of interest to the medical microbiologist, written by specialists for both specialists and non-specialists.

Most *Digest* articles are written by PHLS staff — from the Central Public Health Laboratory, the Communicable Disease Surveillance Centre, the Centre for Applied Microbiology and Research, and public health laboratories.

Recent issues of the journal have covered a wide range of topics including plasmid reference, microbiological standards for raw milk, *Clostridium difficile* diagnosis, smallpox virus and its experimental alternatives, genetic engineering, *Staphylococcus aureus* outbreaks in hospitals, getting papers published in scientific journals, dental caries, bedside diagnostic techniques, protozoal diarrhoea, the pyogenic streptococci, holiday infections, cryptosporidiosis, paediatric epidemiology and hepatitis B among the mentally handicapped. These articles are supplemented by briefer reports on developments in public health, international meetings, etc.

The April 1986 issue was devoted entirely to salmonella. Six articles reviewed infection trends, antibiotic resistance, veterinary aspects, milk and dairy products, irradiation processing, and food hygiene. Another special issue, on vaccination, was published in April 1987; it included meningitis, AIDS, hepatitis B, rubella, pertussis and cytomegalovirus.

PHLS Library Bulletin

ISSN 0267 6850, published weekly, two volumes per annum, approximately 600 A4 pages and two indexes per annum. Subscriptions (1989): UK individuals, £25; UK institutions, £45; overseas individuals, £45; overseas institutions £70.

PHLS Library Bulletin is a current awareness bulletin compiled specifically for the public health and medical laboratory microbiologist. Each week it lists the authors and titles of relevant papers published in over 400 of the world's journals — from the *Lancet* to the *Islamic World Medical Journal* and from the *Journal of Diarrhoeal Diseases Research* to the *Journal of Biological Chemistry.* The Bulletin contains some 350 citations per issue.

The *Bulletin* was started for internal use in the PHLS in 1949. Staff continue to find it invaluable for minimising their literature searches. Since 1984 outside subscribers have been able to enjoy the same benefits.

PHLS books

Public health virology
12 reports

Edited by Philip P Mortimer, published 1986, ISBN 0 901144 20 7, 193 pages, £15.

Viruses are the cause of most transmissible disease in the community, and developments in their diagnosis, surveillance and control are a matter of keen interest to microbiologists and others concerned with the public health. This book is a collection of reports on medical virology that places particular emphasis on new techniques, new viruses and new issues in the field of public health. It reaches beyond the welter of new information presented in journals and symposia to identify some of the issues that bear on work in diagnostic laboratories.

All the authors of these reports are, or have been, closely associated with the Public Health Laboratory Service. An important role of the Service is to bring to the community the benefits of advances in the biological sciences, in the shape of new diagnostic procedures, improved surveillance and better prophylaxis against infectious disease. The book is edited by Philip P Mortimer, now Director of the PHLS Virus Reference Laboratory.

Contents Introduction by David AJ Tyrrell, Diagnosis by viral genome detection, Does rubella reinfection matter? Can congenital rubella be eliminated? 40 years of virology in the PHLS, Orphan viruses, orphan diseases, *Mycoplasma hominis* in human disease, Viral diagnosis by antibody capture assay, The contribution of the electron microscope, Human parvovirus B19 and fifth disease, Chlamydial respiratory infection, Hepatitis B in mothers and babies, Thinking about the AIDS epidemic. Includes many high-quality electron micrographs and other illustrations.

Yersiniosis
Lab diagnosis Clinical features Epidemiology

By Nicholas S Mair and Eric Fox, published 1986, ISBN 0 901444 16 9, 47 pages, price £10.

This book is concerned primarily with the laboratory investigation of yersiniosis, but also places some emphasis on its clinical and epidemiological aspects, because although the major clinical features of infection with *Y pseudotuberculosis* and *Y enterocolitica* are similar in many respects, there are differences in their frequency, age and sex-related attack rates and epidemiology. Biochemically, *Y pseudotuberculosis* is a remarkably homogenous species whereas *Y enterocolitica* is biochemically heterogeneous, comprising five distinct biovars. The immunological and histological aspects of yersiniosis also differ according to the infecting organism. An awareness of these differences may assist in the proper selection and interpretation of laboratory tests. The text also takes into account the related species which may cause confusion in identification, particularly when they are isolated from human sources.

Contents Clinical features of yersiniosis, The lab diagnosis of human infection, Isolation of *Yersinia* from animals, food and the environment, Identification of *Yersinia*, pathogenicity and virulence, Media details.

Campylobacter III

Edited by AD Pearson, MB Skirrow, H Lior and B Rowe, published 1985, ISBN 0 901144 17 7, 338 pages, price £25.

This volume represents the proceedings of the Third International Workshop on Campylobacter Infections, held in Ottawa in July 1985. Authors' abstracts (totalling 161), updated and edited, are combined with editorial comment on the seven main subject areas. This provides an effective overview of the current knowledge on campylobacter infection. There is also an extensive, two-year bibliography of the literature on campylobacter. The volume is addressed to microbiologists and epidemiologists, but also to taxonomists, immunologists and infectious disease physicians.

Contents Introduction, Historical background, Session I Molecular biology, plasmids and antibiotic resistance, Session II Isolation, detection, serodiagnosis and clinical, Session III Veterinary aspects and food, Session IV Pathogenesis and toxins, Session V Campylobacter-like organisms, Session VI Serotyping and biotyping, Session VII Epidemiology, developing countries and prevention, Bibliography 1983-1985, Author index.

Campylobacter (Reading Workshop, 1981) and *Campylobacter II* (Brussels, 1983) are also still in print, at £8 each, £14 the pair. The three together cost £34.

Microbiology for the Public Health
The evolution of the Public Health Laboratory Service 1939-1980

By REO Williams, published 1985, ISBN 0 901 14415 0 (hardback) 0 901 14418 5 (softback), 223 pages, price £15 (hardback) £10 (softback).

Written by a former Director of the PHLS and dedicated to another, Sir Graham Wilson, on his 90th birthday, *Microbiology for the Public Health* describes the growth and development of the Public Health Laboratory Service from its origins in the 1930s to 1980. It emphasises the aims and objectives during the period as well as medical and scientific successes and failures.

Sir Robert Williams' narrative spans a period during which there were great changes in the prevalent infectious diseases and in both the administrative and technical methods used for their control, with on the one hand the introduction of the National Health Service and on the other an increasing sophistication of technical procedures. As such it is of great interest to people who require an understanding of medical microbiology and public health medicine in the modern world; those directly involved in patient care; routine users and practitioners of medical laboratory services; medical administrators and policy-makers; and of course historians of modern medicine.

Contents Origins in the 1930s, The Emergency Public Health Laboratory Service in wartime, Scientific work of the Emergency Public Health Laboratory Service, Administrative development, The Public Health Laboratory Service: the first decade, The work of the Service 1946-56, Administrative reorganisation: creation of the statutory Public Health Laboratory Service Board, The second phase: 1957-69, Scientific work of the Service, 1957-69, The third phase: 1970-80, The work of the Service 1970-80, The infections in the 1970s, The Centre for Applied Microbiology and Research, Public health laboratory services in Europe, Forty years of microbiology for the public health, Appendix, References, Index. Includes many archive photographs and other illustrations.

Chemical Disinfection in Hospitals

By GAJ Ayliffe, D Coates and PN Hoffman, published 1984, ISBN 0 901 14414 2, 36 pages, price £4.50.

The past decade has seen an important growth in our understanding of the way in which chemical disinfectants work and how they should be used in hospitals to minimise the dangers of cross-infection. The expansion in the use of disposable products and the impact of such items as electronic thermometers have also resulted in procedural changes which also affect requirements for disinfection.

Chemical Disinfection in Hospitals is an important book for all hospital staff concerned with the transfer of infection and the disinfection of facilities and equipment, staff and patients. It is of particular value when drawing up codes of practice for the use of chemical disinfectants in hospitals.

Contents Principles of chemical disinfection, Properties of chemical disinfectants, Organisms of special significance, Disinfection policy, Cleaning and disinfection of the environment, skin and equipment, Use of disinfectants in the laboratory, Use of disinfectants in the post-mortem room, Appendices: testing of disinfectants, cleaning regimes, Selected bibliography.

SEXUALLY TRANSMITTED DISEASES
A rational approach to their diagnosis

Edited by A E Jephcott, published 1987. ISBN 0 901144 21 5, 87 pages, price £10.

We have seen an explosion in the incidence of some sexually transmitted diseases, a degree of containment in others; dramatically, the appearance of a new one—AIDS. But however the pendulum swings, the necessity for good laboratory support remains.

The traditional VD laboratory is no more—but it is often convenient to consider sexually transmitted diseases as a group. This book does so and is, above all, a practical guide to current techniques. The rational approach of its seven chapters gives laboratories the information necessary to select and perform the appropriate tests and interpret results; reference and specialist facilities are also dealt with. Risks to health care staff are discussed.

Contents Syphilis, Gonorrhoea, *Chlamydia trachomatis* infections, Anaerobic vaginosis, Genital herpes, Hepatitis and retroviruses including HIV, Chancroid.

ANAEROBIC INFECTIONS
Clinical and laboratory practice

By A Trevor Willis and Kenneth D Phillips, published 1988, ISBN 0 901144 22 3, 169 pages, price £20.

Anaerobic bacteria cause a variety of disease processes in man – including wound and surgical infections such as tetanus and gas gangrene, botulism and *Clostridium perfringens* food poisoning, and non-clostridial infections like abscess, bacteraemia and empyema.

This book reflects the developments in clinical anaerobic bacteriology and taxonomy of the last decade. The chapters on methodology emphasise accuracy and convenience for day-to-day use in a busy acute clinical laboratory. There are nine chapters, 30 tables, 23 figures and over 350 references.

Contents Infections with obligate anaerobes, Anaerobes as normal flora, Botulism: *Clostridium perfringens* food poisoning; enteritis necroticans, *Clostridium difficile* antibiotic-associated diarrhoea and colitis, Methods of culturing anaerobes, Examination and identification of anaerobes, Diagnosis of anaerobic infections, Anaerobic genera of clinical importance, Gas-liquid chromatography. Includes 23 figures, 30 tables, 350 references.

Trevor Willis PhD DSc MD FRCPA FRACP FRCPath is Director of the PHLS Laboratory at Luton
Ken Phillips PhD is Principal Microbiologist of the PHLS Anaerobe Reference Unit at Luton Public Health Laboratory

To be published in 1989

ELISA
Enzyme-linked immunosorbent assays in the clinical microbiology laboratory

Edited by Dr TG Wreghitt and Dr P Morgan-Capner (Cambridge and Preston Public Health Laboratories).

In the 15 years or so since ELISAS were first introduced as quantitative diagnostic tests, they have become widely used by bacteriologists and virologists. Although they have many advantages over other tests, however, they are not always simple to develop or to perform reliably. This book gives a comprehensive account of the advantages, disadvantages, limitations and interpretation of results of many ELISAS.

The book's many distinguished authors — drawn from outside the PHLS as well as within — evaluate the merits of ELISAS and other methods in diagnosing a wide range of infections in the clinical microbiology laboratory.

Contents Introduction, Techniques, Conjugation, Quality control and standardization, Rubella, Herpes viruses, Respiratory viruses, Coxsackie B virus, Rotavirus, *Clostridium perfringens* enterotoxin etc, Diphtheria and tetanus antitoxins, Campylobacter, Brucellosis, Legionella, Sexually transmitted infections, *Mycoplasma pneumoniae*, *Toxoplasma gondii*, Leptospira, Mycology, Parasitology, New techniques.
Approx 200 pages.
Approx £20.

PUBLIC HEALTH VIROLOGY II
14 reports

A sequel to the successful 1986 book: edited this time by Dr Peter Morgan-Capner, Consultant Virologist at the PHLS laboratory at Preston. Once again, a collection of reports on medical virology that places particular emphasis on newly current issues.

Contents HTLV1 and HTLV2 in the UK, Virology in the National Blood Transfusion Service, Mumps – worthy of elimination? Varicella/zoster in pregnancy, Chronic Q fever, *Mycoplasma pneumoniae*, Quality assessment in virology, Approaches to the diagnosis of *Chlamydia trachomatis* infection, Norwalk virus disease, Non-A, non-B hepatitis, Epstein-Barr virus serology and its application, Water and viruses, Treatment of cytomegalovirus infections, Rabies revisited.

Approximately 200 pages, £20.